PARTIAL DIFFERENTIA
FOR SCIENTISTS AND

(formerly AN INTRODUCTION TO PARTIAL DIFFERENTIAL EQUATIONS FOR SCIENCE STUDENTS)

Partial Differential Equations for Scientists and Engineers

G. STEPHENSON
B.Sc., Ph.D., D.I.C.
Emeritus Reader in Mathematics, Imperial College, London

LONGMAN
London and New York

LONGMAN GROUP LTD
Longman House, Burnt Mill, Harlow
Essex CM20 2JE England

Associated companies throughout the world

*Published in the United States of America
by Longman Inc., New York*

First published 1968
Second edition 1970
*New impression, with
minor amendments
1974*
Reprinted 1978, 1980
Third edition 1985
Reprinted 1986

British Library Cataloguing in Publication Data

Stephenson, G.
  *Partial differential equations for scientists
  and engineers. — 3rd ed.*
  1. Differential equations. Partial
  I. Title II. Stephenson, G. Introduction to
  partial differential equations for science
  students
  515.3'53 QA374

ISBN   0-582-44696-1

*Produced by Longman Group (FE) Ltd
Printed in Hong Kong*

CONTENTS

Contents

Contents

10. The Numerical Approach

PREFACE

ALTHOUGH numerous elementary and inexpensive books on ordinary differential equations are available for students of science and engineering, there are few treating partial differential equations at the same level. This book is an attempt to meet this need and gives an introduction to some of the main methods of solution of linear partial differential equations.

As with the author's *An Introduction to Matrices, Sets and Groups for Science Students* (Longmans, 1965), this companion text is written primarily for second and third year undergraduate physicists, chemists and engineers. A basic first year ancillary mathematics course is therefore assumed, but beyond this the treatment is virtually self-contained. The subject of partial differential equations, however, is so wide and involves so many different branches of mathematics that some careful selection of material has proved necessary. Accordingly we have concentrated almost entirely on second order linear equations of the type that arise in the simpler problems of science and engineering. These are the so-called equations of mathematical physics and include, amongst others, the wave-equation, the heat conduction equation, Laplace's equation and Schrödinger's equation. The emphasis of the treatment here is on the methods of solution rather than on the solutions themselves, and consequently little space is devoted to the special functions (e.g. Bessel functions, Legendre polynomials, etc.) which can arise in certain types of problems. Two main methods of solution are discussed; namely, Fourier's method based on the separation of variables technique, and the method of integral transforms. In the last chapter a brief look is taken at other methods which, for reasons of space, cannot be dealt with in detail here.

No attempt is made to enter into the various numerical methods of solution, important as they are. The reader who wishes to obtain an elementary introduction to these aspects should consult the text referred to in the list of Further Reading given at the end of this book.

The author wishes to thank Dr. I. N. Baker, Dr. A. N. Gordon and Professor C. W. Kilmister for reading the manuscript and making various criticisms and suggestions which have improved the text. Especial thanks are due to Dr. Gordon for reading the proofs and making a number of pertinent comments.

Imperial College, London.   G. S.
1967

PREFACE TO THE SECOND EDITION

In this new edition an extra chapter has been included dealing more fully with the method of Green's functions for the solution of inhomogeneous equations.

Topics which relate to this method and which play an important part in the more advanced theory of partial differential equations, such as the Dirac delta function and other generalised functions, are introduced in an elementary way.

The opportunity has also been taken to correct a number of minor misprints which occurred in the first edition, and the author is grateful to many friends and colleagues for pointing these out.

Imperial College, London.   G. S.
1970

PREFACE TO THE THIRD EDITION

In this third edition Chapter 10 has been replaced by new material covering the basic elements of some numerical methods for the solution of differential equations. This is an important field of study and no student can now afford to be unacquainted with the numerical approach to the subject. In the space available it is not possible to do justice to the variety of techniques which exist, and accordingly only two methods are discussed in order to give the flavour of the subject. These are (a) the finite difference method and (b) the finite element method. For more detailed and substantial accounts of numerical methods the reader is referred to the appropriate texts listed under Further Reading at the end of this book.

Finally, the minor change in the title emphasises the importance of partial differential equations to engineering students as well as to physical scientists.

Imperial College, London.   G. S.
1984

CHAPTER 1

Basic Concepts

1.1 Introduction

In this introductory chapter we shall be concerned principally with the occurrence and nature of certain types of partial differential equations leaving aside the discussion of particular methods of solution to the remaining chapters. Moreover, the types of equations discussed here and in later parts of this book are primarily those which have their origins in the mathematical description of physical processes and laws. Firstly, however, we consider some general ideas and definitions.

Any differential equation containing partial derivatives is called a partial differential equation, the order of the equation being equal (by analogy with the theory of ordinary differential equations) to the order of the highest partial differential coefficient occurring in it. The dependent variable (the unknown function) in any partial differential equation must be a function of at least two independent variables otherwise partial derivatives would not arise, and in general may be a function of $n(\geqslant 2)$ independent variables. For example, the equations

$$3y^2 \frac{\partial u}{\partial x} + \frac{\partial u}{\partial y} = 2u, \tag{1}$$

$$\frac{\partial^2 u}{\partial x^2} + f(x, y) \frac{\partial^2 u}{\partial y^2} = 0, \tag{2}$$

(where $f(x, y)$ is any given function) are typical partial differential equations of the first and second orders respectively, x and y being the independent variables, and $u \equiv u(x, y)$ being the dependent variable whose form is to be found by solving the appropriate equation.

Likewise the equation

$$\frac{\partial^2 u}{\partial x^2} + \frac{\partial^2 u}{\partial y^2} + \frac{\partial^2 u}{\partial z^2} = 0 \tag{3}$$

(known as Laplace's equation in three variables) is an equation for $u(x, y, z)$ where x, y and z are independent variables. We note that

1

equations (1), (2) and (3) are all linear in the sense that u and its partial derivatives occur only to the first degree, and that products of u and its derivatives are absent. A typical non-linear equation in two independent variables is

$$u \frac{\partial^2 u}{\partial x^2} + \left(\frac{\partial u}{\partial y}\right)^2 = u^2. \tag{4}$$

The solution of non-linear partial differential equations will not be considered in this book. In general, the solution of partial differential equations presents a much more difficult problem than the solution of ordinary differential equations, and except for certain special types of *linear* partial differential equations, no general method of solution is available. We shall concentrate therefore on the solution of particular types of linear equations. This is in fact not too serious a restriction since linear partial differential equations have a great variety of important applications in many branches of physics, chemistry and engineering. To understand the occurrence of partial differential equations in the mathematical description of the phenomena of Nature we note that most physical processes or events are described by functions of two or more (usually four) independent variables – for example, one, two or three space variables, x, y and z, and a time variable t. Consequently any relation between such a function ($u(x, y, z, t)$, say) and its derivatives with respect to any of the independent variables will lead to a partial differential equation. Much of the mathematical study of partial differential equations has been directed towards an understanding of that class of equations known popularly as the partial differential equations of mathematical physics. The field of mathematical physics in this context is to be interpreted in the widest sense – that is, the description of the phenomena of Nature in mathematical terms. Included in this class of equations, therefore, are not only those equations of importance in modern theoretical physics (such as the Schrödinger and Dirac equations of quantum theory) but those equations of importance to the applied mathematician and engineer (for example, the heat conduction or diffusion equation, the equations of viscous fluid flow, and many others). It is clearly almost impossible in these days of interplay between different branches of science and technology to single out particular equations as being of little importance to particular types of scientists. In any event, the same equation often turns up in a variety of different physical situations as we shall see in 1.2. However, it is both remarkable and fortunate that a large number of the important equations in mathematical physics are not only linear but of second-order. This is not to say, however, that other types of equations do not arise. For example, the Dirac equation of quantum theory is linear but of first-order, whilst the equations of general relativity which describe the gravitational field are second-order but non-linear. Similarly an equation of importance in elasticity (the bi-harmonic equation) is linear but of fourth-order. The complexity of the solution of a linear equation, besides depending on the order of the equation, is strongly dependent on the number of independent variables involved, and accordingly in many cases we illustrate a particular method of solution by first taking as examples partial differential equations in two independent variables. These equations are in a sense intermediate in mathematical difficulty between ordinary differential equations on the one hand and partial differential equations in three or more independent variables on the other.

A linear equation is said to be *homogeneous* if each term contains either the dependent variable or one of its derivatives. For example, Laplace's equation in two-dimensions (that is, two independent variables)

$$\nabla^2 u = 0, \tag{5}$$

where ∇^2 is the two-dimensional Laplace operator (defined in rectangular Cartesian coordinates (x, y) by $\nabla^2 = \dfrac{\partial^2}{\partial x^2} + \dfrac{\partial^2}{\partial y^2}$) is homogeneous. However, the two-dimensional Poisson equation

$$\nabla^2 u = f(x, y), \tag{6}$$

where $f(x, y)$ is any given (non-zero) function, is termed inhomogeneous (or non-homogeneous). Now in the case of linear homogeneous ordinary differential equations it is well known that a linear combination of two or more solutions is also a solution. A similar result applies to partial differential equations, and if $u_1, u_2, \ldots, u_n$, where n may be finite or non-finite, are n different solutions of a linear homogeneous partial differential equation in some given domain then

$$u = c_1 u_1 + c_2 u_2 + \ldots + c_n u_n \tag{7}$$

3

is also a solution in the same domain, where the coefficients c_1, c_2, ..., c_n are arbitrary constants. This result is known as the Principle of Superposition and has an important place in the method of solution known as the separation of variables (see Chapter 2).

We now come to one of the most important differences between the solutions of partial differential equations and those of ordinary differential equations. For whereas the general solution of a linear ordinary differential equation contains arbitrary constants of integration, the general solution of a linear partial differential equation contains arbitrary functions. To illustrate this point we consider the problem of the formation of partial differential equations from given functions. For example, if

$$u = y f(x), \tag{8}$$

where $f(x)$ is an arbitrary function of x, then differentiating with respect to y we have

$$\frac{\partial u}{\partial y} = f(x). \tag{9}$$

Eliminating $f(x)$ between (8) and (9) we find

$$y \frac{\partial u}{\partial y} = u, \tag{10}$$

which is a first-order linear partial differential whose general solution is given by (8). The significant point here is that the solution of (10) as given by (8) contains an arbitrary function.

Similarly if

$$u = f(x + y) + g(x - y), \tag{11}$$

where $f(x + y)$ and $g(x - y)$ are arbitrary functions of $x + y$ and $x - y$ respectively, then

$$\frac{\partial u}{\partial x} = f'(x + y) + g'(x - y), \tag{12}$$

$$\frac{\partial^2 u}{\partial x^2} = f''(x + y) + g''(x - y), \tag{13}$$

$$\frac{\partial u}{\partial y} = f'(x + y) - g'(x - y), \tag{14}$$

$$\frac{\partial^2 u}{\partial y^2} = f''(x + y) + g''(x - y), \tag{15}$$

where dashes denote differentiation with respect to the appropriate argument $(x+y$ or $x-y)$. Hence by equating (13) and (15) and so eliminating the arbitrary functions we obtain the second-order partial differential equation

$$\frac{\partial^2 u}{\partial x^2} = \frac{\partial^2 u}{\partial y^2}. \qquad (16)$$

The function u defined by (11) therefore satisfies (16) irrespective of the functional forms of $f(x+y)$ and $g(x-y)$, provided f and g are at least twice differentiable functions. For example,

$$u = \sin(x+y) + e^{x-y}, \text{ and } u = (x+y)^3 + \tan(x-y), \qquad (17)$$

are both solutions of (16). As with the previous example (eqs. (8)–(10)), the general solution of (16) as given by (11) contains arbitrary functions.

In most cases the general solution of a partial differential equation is of little use since it has to be made to satisfy other conditions – called boundary conditions – which arise from the physics of the problem. This is much more difficult to accomplish for partial differential equations than for ordinary differential equations owing to the great variety of choice available for the arbitrary functions. For this reason (and also for the reason that general solutions are not often known) other methods of solution such as the methods of separation of variables and integral transforms, which build the solution of an equation around the boundary conditions, are of great importance. These methods are discussed in detail in the remaining chapters of this book.

Finally, we should mention here that the term 'boundary condition' is used in the literature with different meanings. The term is clearly appropriate when an equation has to be solved within a given region R of *space*, with prescribed values of the dependent variable given on the boundary of R. Moreover, the boundary need not enclose a finite volume – in which case part of the boundary will be at infinity. In the case of partial differential equations in which one of the independent variables is the time t, the values of the dependent variable and often its time derivative at some instant of time, say $t=0$, may be given. Such conditions are usually called 'initial conditions'. Initial conditions, however, may be thought of as boundary conditions in a space-time diagram where one of the axes represents the time coordinate. In the case of a partial differential equation in two independent variables, say a space variable x and a time variable t, we might require a solution within the region R (see Fig. 1.1). Here the initial conditions at $t=0$ are boundary conditions along the boundary OA.

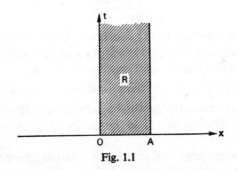

Fig. 1.1

In this book we shall use the terms 'boundary condition' and 'initial condition' as appropriate, particularly in 1.5 where a more detailed discussion of boundary conditions is given.

1.2 The wave equation

As we have remarked in the previous section, many of the important equations of mathematical physics are linear second-order partial differential equations. Many of these can be derived from first principles after a detailed analysis of the physical situation. We shall not concern ourselves with the derivations of the various types of equations, but the following analysis relating to the one-dimensional wave equation is included as an example of how a partial differential equation may be set up to represent a physical process. It should be pointed out here, however, that not all the equations of importance in mathematical physics are derivable in this way and that some (for example, the Schrödinger and Dirac equations of quantum theory, and the equations of general relativity) are essentially *postulated* as axioms of a theory.

Consider a perfectly flexible string of uniform density ρ stretched to a uniform tension T between two points $x=0$ and $x=l$ (see Fig. 1.2). Now since, by assumption, the string offers no resistance to bending the tension is tangential to the string at each point. If we let T_1 and T_2 be the tensions at the points A and B respectively, then

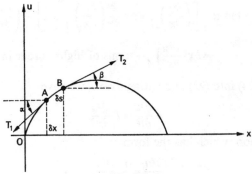

Fig. 1.2

since all the motion is transverse (i.e. perpendicular to the x-axis) we must have
$$T_1 \cos \alpha = T_2 \cos \beta = T(= \text{constant}). \tag{18}$$

Resolving in the u-direction, we have for the small element AB
$$\rho \, \delta s \, \frac{\partial^2 u}{\partial t^2} = T_2 \sin \beta - T_1 \sin \alpha, \tag{19}$$

where δs is the length of AB.

Hence, using (18),
$$\rho \, \frac{\delta s}{T} \, \frac{\partial^2 u}{\partial t^2} = \frac{T_2 \sin \beta}{T_2 \cos \beta} - \frac{T_1 \sin \alpha}{T_1 \cos \alpha} = \tan \beta - \tan \alpha. \tag{20}$$

Now assuming small transverse displacements of the string, we may write to a first approximation
$$\delta s = \delta x. \tag{21}$$

Hence
$$\rho \, \frac{\delta x}{T} \, \frac{\partial^2 u}{\partial t^2} = \tan \beta - \tan \alpha. \tag{22}$$

But $\tan \alpha$ and $\tan \beta$ are the gradients of the string at the points A and B respectively, and as such are given by
$$\tan \alpha = \left(\frac{\partial u}{\partial x}\right)_A, \quad \tan \beta = \left(\frac{\partial u}{\partial x}\right)_B, \tag{23}$$

the *partial* derivatives being required since u is a function of both x and t. Since A and B are separated only by a small distance we may

expand $\left(\dfrac{\partial u}{\partial x}\right)_B$ to get

$$\tan \beta - \tan \alpha = \left[\left(\frac{\partial u}{\partial x}\right)_A + \delta x \frac{\partial}{\partial x}\left(\frac{\partial u}{\partial x}\right) + \ldots \right] - \left(\frac{\partial u}{\partial x}\right)_A \quad (24)$$

$$= \delta x \left(\frac{\partial^2 u}{\partial x^2}\right)_A + \text{terms of higher order in } \delta x.$$

Inserting (24) into (22) and letting $\delta x \to 0$, we have

$$\rho \frac{\partial^2 u}{\partial t^2} = T \frac{\partial^2 u}{\partial x^2}. \quad (25)$$

This equation, which has the form

$$\frac{\partial^2 u}{\partial x^2} = \frac{1}{c^2} \frac{\partial^2 u}{\partial t^2}, \quad (26)$$

where $c^2 = T/\rho$, is called the one-dimensional wave equation and describes the motion of the string when the displacements are small. For large transverse displacements the equation is no longer valid, and a mathematical analysis of the situation leads to a non-linear partial differential equation. Consequently (26) does not give an exact mathematical description of the physical process. Moreover the one-dimensional wave equation derived here is only an approximation to the physics of the problem for yet another reason – that is, that the mathematical string is a *continuous* distribution of mass whereas the physical string is composed of a large (but finite) number of individual particles (atoms and molecules) separated by very small but nevertheless finite distances, and held together by atomic and molecular forces. Such a system of particles possesses a finite number of degrees of freedom, whereas the continuous distribution possesses an infinity. The analysis of a system of finite particles attached at equal intervals to a weightless string and performing small transverse oscillations is more analogous to the physical picture and shows clearly how the one-dimensional wave equations (26) is but an idealisation.

Suppose we have N particles, each of mass m, strung together at equal intervals h on a massless string. Let u_j be the transverse displacement of the jth particle, and assume that all displacements are small enough that the tension T of the string is constant (see Fig. 1.3). Then if θ_j is the angle between the line joining the jth and $(j+1)^{th}$ particle and the x-axis, we have (since the displacements are small)

$$\sin \theta_j \simeq \tan \theta = \frac{u_{j+1} - u_j}{h}. \quad (27)$$

8

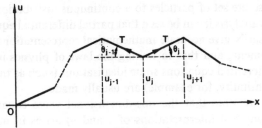

Fig. 1.3

Now the components of force acting on the jth particle are

$$T \sin \theta_{j-1} \simeq \frac{T}{h} (u_{j-1} - u_j) \qquad (28)$$

and

$$T \sin \theta_j \simeq \frac{T}{h} (u_{j+1} - u_j). \qquad (29)$$

Hence the equation of motion of the jth particle is

$$m \frac{\partial^2 u_j}{\partial t^2} = T \sin \theta_j + T \sin \theta_{j-1} \simeq \frac{T}{h} (u_{j+1} - 2u_j + u_{j-1}). \qquad (30)$$

However, by Taylor's series,

$$u_{j+1} = u_j + h \left(\frac{\partial u_j}{\partial x} \right) + \frac{h^2}{2!} \left(\frac{\partial^2 u_j}{\partial x^2} \right) + \dots \qquad (31)$$

$$u_{j-1} = u_j - h \left(\frac{\partial u_j}{\partial x} \right) + \frac{h^2}{2!} \left(\frac{\partial^2 u_j}{\partial x^2} \right) - \dots \qquad (32)$$

and hence

$$\frac{\partial^2 u_j}{\partial x^2} = \lim_{h \to 0} \left(\frac{u_{j+1} - 2u_j + u_{j-1}}{h^2} \right). \qquad (33)$$

Consequently if we now allow the separation between the particles to tend to zero so that the number of particles per unit length becomes increasingly larger and the mass of each particle correspondingly smaller, we obtain in the limit a continuous mass distribution with density ρ, say, whose equation of motion is (using (30) and (33))

$$\rho \frac{\partial^2 u}{\partial t^2} = T \frac{\partial^2 u}{\partial x^2}. \qquad (34)$$

This is the one-dimensional wave equation (26) derived by passing

9

from a discrete set of particles to a continuous distribution of mass. From this analysis it can be seen that partial differential equations do not necessarily give accurate mathematical representations of physical phenomena, and that in writing the laws of physics in terms of partial differential equations some idealisations (such as the assumption of continuity, for example) are usually made.

Finally we note that the one-dimensional wave equation (with different physical interpretations of c and u) arises in many other problems such as the longitudinal vibration of a bar, the propagation of sound waves, and the transmission of electric signals along a cable, to name but a few. The two- and three-dimensional wave equations which take the general form

$$\nabla^2 u = \frac{1}{c^2} \frac{\partial^2 u}{\partial t^2}, \quad (c = \text{constant}) \tag{35}$$

where ∇^2 is the Laplacian operator in two or three independent variables, likewise arise in the description of a variety of physical phenomena. In rectangular Cartesian coordinates (x, y, z) the form of this Laplacian operator is

$$\frac{\partial^2}{\partial x^2} + \frac{\partial^2}{\partial y^2} + \frac{\partial^2}{\partial z^2}, \tag{36}$$

with the obvious omission of one of these terms for the two-dimensional case. However, as we shall see later on, rectangular Cartesian coordinates are not always the most appropriate coordinates to use in a particular problem and accordingly the forms of ∇^2 in other coordinate systems (such as spherical polar coordinates) are of importance. These will be discussed in more detail in Chapter 2.

1.3 Some important equations

In this section we list a few of the simple equations of importance in mathematical physics. Such a list naturally cannot ever be comprehensive, and in any event we have included here only linear equations, whereas many physical phenomena (such as viscous fluid flow) are described by non-linear equations. However, these simpler linear equations form a basis for the understanding of a large number of physical processes and furthermore are soluble by a variety of important mathematical techniques, some of which are described in the remaining chapters of this book.

10

Besides the wave equation

$$\nabla^2 u = \frac{1}{c^2} \frac{\partial^2 u}{\partial t^2} \tag{37}$$

described in the last section, we have

(a) $\nabla^2 u = \frac{1}{k} \frac{\partial u}{\partial t}$ (the heat conduction or diffusion equation). (38)

(b) $\nabla^2 u = 0$ (Laplace's equation). (39)

(c) $\nabla^2 u + \lambda u = 0$ (Helmholtz's equation). (40)

(d) $\nabla^2 u = f(x, y, z)$ (Poisson's equation). (41)

(e) $\nabla^4 u = \nabla^2(\nabla^2 u) = -\frac{1}{p^2} \frac{\partial^2 u}{\partial t^2}$ (the biharmonic wave equation). (42)

(f) $\nabla^4 u = 0$ (the biharmonic equation). (43)

(g) $\nabla^2 u + \alpha[E - V(x, y, z)]u = 0$ (Schrödinger's equation). (44)

(h) $\square u + \lambda^2 u = 0$ (the Klein-Gordon equation) (45)
where the operator $\square$, called the D'Alembertian operator, is defined as

$$\square \equiv \nabla^2 - \frac{1}{c^2} \frac{\partial^2}{\partial t^2}. \tag{46}$$

In all these equations ∇^2 is the Laplacian operator in the appropriate number of space dimensions, t is the time variable, and $c, k, \lambda, p, \alpha, E$ are constants whose physical interpretations (like that of the dependent variable u) are dictated by the particular problem on hand. The functions f and V are usually assumed to be known.

As with ordinary differential equations, the solutions of partial differential equations are usually required to satisfy certain constraints called boundary conditions. The types of boundary conditions which arise in some of the simpler problems of mathematical physics will be discussed in Chapter 2. For the moment, however, we remark that in the cases of Helmholtz's equation (c) and Schrödinger's equation (g) the constants λ and E respectively have to take on special values (the eigenvalues, as they are called) in order that the boundary conditions may be satisfied. Usually these eigenvalues form an infinite discrete set of values, and to each eigenvalue there is a corresponding solution u called an eigenfunction. Examples of eigenvalue-eigenfunction problems will be met in Chapters 3–5.

11

Finally we note that the biharmonic equation (f) possesses all the solutions of Laplace's equations but in addition has an infinity of others. For example, it is easily verified that for any four times differentiable function $u(x, y)$

$$\nabla^4(xu) = x\nabla^4 u + 4\frac{\partial}{\partial x}(\nabla^2 u) \tag{47}$$

and

$$\nabla^4(yu) = y\nabla^4 u + 4\frac{\partial}{\partial y}(\nabla^2 u). \tag{48}$$

Hence if u is a solution of Laplace's equation (that is, $\nabla^2 u = 0$) then both xu and yu satisfy the biharmonic equation, but, in general, will not satisfy Laplace's equation.

PROBLEMS 1

1. Eliminate the arbitrary functions from the following functions, so obtaining partial differential equations of which the general solutions are

   (a) $u = f(x+y)$,   (b) $u = f(xy)$,
   (c) $u = f(x+y) + g(x-y)$,   (d) $u = x^n f(y/x)$.

2. Show that

   $$u(x, y, t) = f(x + i\beta y - vt) + g(x - i\beta y - vt)$$

   is a solution of the wave-equation

   $$\frac{\partial^2 u}{\partial x^2} + \frac{\partial^2 u}{\partial y^2} = \frac{1}{c^2}\frac{\partial^2 u}{\partial t^2},$$

   where f and g are arbitrary (twice differentiable) functions, and

   $$\beta = \sqrt{\left(1 - \frac{v^2}{c^2}\right)},$$

   β, v, c, being constants.

3. Show that

   $$u = f(2x + y^2) + g(2x - y^2)$$

   satisfies the equation

   $$y^2\frac{\partial^2 u}{\partial x^2} + \frac{1}{y}\frac{\partial u}{\partial y} - \frac{\partial^2 u}{\partial y^2} = 0,$$

   where f and g are arbitrary (twice differentiable) functions.

12

CHAPTER 2

Classification of Equations and Boundary Conditions

2.1 Types of equations

In the previous chapter a number of physically important equations were briefly discussed. Many of these equations, when only two independent variables are present, are special cases of the general linear homogeneous equation of the second-order, namely

$$a\,\frac{\partial^2 u}{\partial x^2} + 2h\,\frac{\partial^2 u}{\partial x \partial y} + b\,\frac{\partial^2 u}{\partial y^2} + 2f\,\frac{\partial u}{\partial x} + 2g\,\frac{\partial u}{\partial y} + eu = 0, \qquad (1).$$

where a, h, b, f, g and e may be constants or functions of x and y. For example, the wave equation

$$\frac{\partial^2 u}{\partial x^2} = \frac{1}{c^2}\,\frac{\partial^2 u}{\partial t^2} \qquad (2)$$

may be obtained from (1) by associating the independent variable y with the time variable t, and choosing $a=1$, $h=0$, $b=-\dfrac{1}{c^2}$, $f=g=e=0$. We shall return to this and other special cases later on. For the moment we note that the form of (1) resembles that of a general conic section

$$ax^2 + 2hxy + by^2 + 2fx + 2gy + e = 0. \qquad (3)$$

This equation represents an ellipse, parabola or hyperbola when $ab - h^2 > 0$, $= 0$, < 0 respectively. We use a similar classification for the partial differential equation (1) and say that it is of

$$\left.\begin{array}{l} \text{elliptic} \\ \text{parabolic} \\ \text{hyperbolic} \end{array}\right\} \text{ type when } \left\{\begin{array}{l} ab - h^2 > 0, \\ ab - h^2 = 0, \\ ab - h^2 < 0. \end{array}\right. \qquad (4)$$

Consequently the wave equation (2) is of hyperbolic type since

$$ab - h^2 = -\frac{1}{c^2} < 0.$$

Laplace's equation in two variables

$$\frac{\partial^2 u}{\partial x^2} + \frac{\partial^2 u}{\partial y^2} = 0 \qquad (5)$$

may be obtained from (1) by putting $a=1$, $h=0$, $b=1$, $f=g=e=0$, and hence, since $ab-h^2=1>0$, is of elliptic type. However, the equation

$$\frac{\partial^2 u}{\partial x^2} = k\,\frac{\partial u}{\partial y} \qquad (6)$$

is of parabolic type since, comparing with (1), $a=1$, $h=0$, $b=0$, $f=0$, $g=-\dfrac{k}{2}$, $e=0$ and $ab-h^2=0$.

Equations in which a, b and h are functions of x and y (variable coefficient equations) may change their type on passing from one region of the xy-plane to another. For example, the equation

$$y\,\frac{\partial^2 u}{\partial x^2} + 2x\,\frac{\partial^2 u}{\partial x \partial y} + y\,\frac{\partial^2 u}{\partial y^2} = 0 \qquad (7)$$

is elliptic in the region where $y^2 - x^2 > 0$, parabolic along the lines $y^2 - x^2 = 0$, and hyperbolic in the region where $y^2 - x^2 < 0$ (see Fig. 2.1).

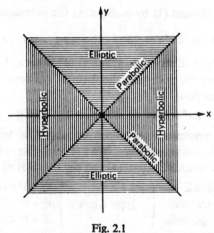

Fig. 2.1

A similar but more complicated classification can be carried out for linear equations in three or more independent variables. In the case

14

of three independent variables the terms elliptic, parabolic and hyperbolic should be replaced by their three-dimensional analogues (ellipsoidal, etc.). However, it is customary to continue to use the two-dimensional terms, and equations such as the wave equation

$$\nabla^2 u = \frac{1}{c^2} \frac{\partial^2 u}{\partial t^2} \qquad (8)$$

whether it be in two, three, or four independent variables is in general called hyperbolic. Likewise Laplace's equation

$$\nabla^2 u = 0 \qquad (9)$$

in two or three independent variables is elliptic in type. Similarly the heat conduction equation

$$\nabla^2 u = \frac{1}{k} \frac{\partial u}{\partial t} \qquad (10)$$

is of parabolic type.

As we shall see in the next section, the forms of the general solutions of linear partial differential equations depend very much on the type (elliptic, parabolic or hyperbolic) of the equation considered. Similarly the forms of the boundary conditions which may be imposed on the three types of equations to yield unique solutions depend also on the type of equation. Boundary conditions will be discussed in greater detail in 2.3.

2.2 Euler's equation

The equation

$$a \frac{\partial^2 u}{\partial x^2} + 2h \frac{\partial^2 u}{\partial x \partial y} + b \frac{\partial^2 u}{\partial y^2} = 0 \qquad (11)$$

where a, h and b are constants, is a special case of (1) and is usually known as Euler's equation. The general solution of this equation may be obtained in the following way.

We first define two new independent variables ξ and η by the linear relations

$$\begin{aligned} \xi &= px + qy, \\ \eta &= rx + sy, \end{aligned} \qquad (12)$$

where p, q, r and s are arbitrary constants. Then

$$\frac{\partial u}{\partial x} = \frac{\partial u}{\partial \xi} \frac{\partial \xi}{\partial x} + \frac{\partial u}{\partial \eta} \frac{\partial \eta}{\partial x} = p \frac{\partial u}{\partial \xi} + r \frac{\partial u}{\partial \eta}, \qquad (13)$$

15

$$\frac{\partial u}{\partial y} = \frac{\partial u}{\partial \xi}\frac{\partial \xi}{\partial y} + \frac{\partial u}{\partial \eta}\frac{\partial \eta}{\partial y} = q\frac{\partial u}{\partial \xi} + s\frac{\partial u}{\partial \eta}, \quad (14)$$

$$\frac{\partial^2 u}{\partial x^2} = \frac{\partial}{\partial x}\left(\frac{\partial u}{\partial x}\right) = \left(p\frac{\partial}{\partial \xi} + r\frac{\partial}{\partial \eta}\right)\left(p\frac{\partial u}{\partial \xi} + r\frac{\partial u}{\partial \eta}\right) \quad (15)$$

$$= p^2\frac{\partial^2 u}{\partial \xi^2} + 2pr\frac{\partial^2 u}{\partial \xi\partial \eta} + r^2\frac{\partial^2 u}{\partial \eta^2}, \quad (16)$$

$$\frac{\partial^2 u}{\partial y^2} = \frac{\partial}{\partial y}\left(\frac{\partial u}{\partial y}\right) = \left(q\frac{\partial}{\partial \xi} + s\frac{\partial}{\partial \eta}\right)\left(q\frac{\partial u}{\partial \xi} + s\frac{\partial u}{\partial \eta}\right) \quad (17)$$

$$= q^2\frac{\partial^2 u}{\partial \xi^2} + 2sq\frac{\partial^2 u}{\partial \xi\partial \eta} + s^2\frac{\partial^2 u}{\partial \eta^2}, \quad (18)$$

and

$$\frac{\partial^2 u}{\partial x\partial y} = \frac{\partial}{\partial x}\left(\frac{\partial u}{\partial y}\right) = \left(p\frac{\partial}{\partial \xi} + r\frac{\partial}{\partial \eta}\right)\left(q\frac{\partial u}{\partial \xi} + s\frac{\partial u}{\partial \eta}\right) \quad (19)$$

$$= pq\frac{\partial^2 u}{\partial \xi^2} + (rq + sp)\frac{\partial^2 u}{\partial \xi\partial \eta} + rs\frac{\partial^2 u}{\partial \eta^2}. \quad (20)$$

Substituting these expressions for the second partial derivatives into (11) we find

$$(ap^2 + 2hpq + bq^2)\frac{\partial^2 u}{\partial \xi^2} + 2[apr + bsq + h(rq + sp)]\frac{\partial^2 u}{\partial \xi\partial \eta} +$$

$$+ (ar^2 + 2hrs + bs^2)\frac{\partial^2 u}{\partial \eta^2} = 0. \quad (21)$$

We now choose arbitrary constants p, q, r and s such that $p = r = 1$ and such that q and s are the two roots λ_1 and λ_2 of the equation

$$a + 2h\lambda + b\lambda^2 = 0. \quad (22)$$

Consequently (21) becomes

$$[a + h(\lambda_1 + \lambda_2) + b\lambda_1\lambda_2]\frac{\partial^2 u}{\partial \xi\partial \eta} = 0. \quad (23)$$

However, since

$$\lambda_1 + \lambda_2 = -\frac{2h}{b}, \quad (24)$$

$$\lambda_1\lambda_2 = \frac{a}{b}, \quad (25)$$

16

(23) may be written as

$$\frac{2}{b}(ab - h^2)\frac{\partial^2 u}{\partial \xi \partial \eta} = 0. \tag{26}$$

Hence, provided (11) is not parabolic (i.e. $ab - h^2 \neq 0$ and $b=0$), (26) may be integrated to give

$$u = F(\xi) + G(\eta), \tag{27}$$

where F and G are arbitrary functions. Consequently, since $\xi = x + \lambda_1 y$, $\eta = x + \lambda_2 y$, the general solution of (11) when it is not of parabolic type is

$$u = F(x + \lambda_1 y) + G(x + \lambda_2 y), \tag{28}$$

where λ_1, λ_2 are the roots of (22).

We now consider the two possible cases.

Case 1. $ab - h^2 < 0$ (hyperbolic equations).
The roots of (22), λ_1, λ_2 are *real* and *distinct*.

Consequently the general solution (28) is the sum of two arbitrary functions of real arguments.

Example 1. The wave equation

$$\frac{\partial^2 u}{\partial x^2} - \frac{1}{c^2}\frac{\partial^2 u}{\partial t^2} = 0 \tag{29}$$

is a special case of (11) obtained by putting $a = 1$, $h = 0$, $b = -\frac{1}{c^2}$ and replacing the y coordinate by t. Consequently since $ab - h^2 < 0$, the equation is hyperbolic. From (22) we have

$$1 - \frac{\lambda^2}{c^2} = 0, \tag{30}$$

whence $\lambda_1 = c$, $\lambda_2 = -c$.

Hence by (28), the general solution of (29) is

$$u = F(x + ct) + G(x - ct). \tag{31}$$

Example 2. The equation

$$2\frac{\partial^2 u}{\partial x^2} + 3\frac{\partial^2 u}{\partial x \partial y} + \frac{\partial^2 u}{\partial y^2} = 0 \tag{32}$$

is hyperbolic since $a = 2$, $b = 1$, $h = 3/2$ and $ab - h^2 < 0$. By (22) we have

$$2 + 3\lambda + \lambda^2 = 0 \tag{33}$$

which gives $\lambda_1 = -1$, $\lambda_2 = -2$.

Hence the general solution of (32) is

$$u = F(x - y) + G(x - 2y). \tag{34}$$

Case 2. $ab - h^2 > 0$ (elliptic equations).

The roots λ_1, λ_2 of (22) are now complex conjugate with $\lambda_1 = p + i\sigma$ (say) $= \lambda_2^*$, the star denoting the complex conjugate. Hence

$$\xi = x + py + i\sigma y,$$

and

$$\eta = x + py - i\sigma y = \xi^*. \tag{35}$$

Equation (26) now becomes

$$\frac{\partial^2 u}{\partial \xi \partial \xi^*} = 0, \tag{36}$$

with the general solution

$$u = F(\xi) + G(\xi^*). \tag{37}$$

The appearance of complex arguments is a general property of the solution of elliptic equations.

Example 3. Laplace's equation in two dimensions

$$\frac{\partial^2 u}{\partial x^2} + \frac{\partial^2 u}{\partial y^2} = 0 \tag{38}$$

is a special case of Euler's equation obtained by putting $a = b = 1$ and $h = 0$. Hence (22) becomes

$$1 + \lambda^2 = 0 \tag{39}$$

giving $\lambda_1 = i$, $\lambda_2 = -i$.

Consequently the general solution is of the form

$$u = F(x + iy) + G(x - iy). \tag{40}$$

Example 4. The equation

$$2\frac{\partial^2 u}{\partial x^2} + \frac{\partial^2 u}{\partial x \partial y} + \frac{\partial^2 u}{\partial y^2} = 0 \tag{41}$$

is elliptic since $ab - h^2 = 2 - (\tfrac{1}{2})^2 > 0$.

Hence (22) becomes

$$2 + \lambda + \lambda^2 = 0 \tag{42}$$

or

$$\lambda = \frac{-1 \pm \sqrt{(1-8)}}{2} = -\frac{1}{2} \pm i\frac{\sqrt{7}}{2}. \tag{43}$$

Consequently

$$u = F\left(x - \frac{1}{2}y + i\frac{\sqrt{7}}{2}y\right) + G\left(x - \frac{1}{2}y - i\frac{\sqrt{7}}{2}y\right) \tag{44}$$

is the general solution.

In deriving (28) we assumed that the equation (11) was not parabolic for in this case $ab - h^2 = 0$ and (26) is then satisfied identically. Under these circumstances we proceed again by making the transformation (12) as before but choosing $p = 1$ and leaving q, r and s arbitrary (for the moment). Then

$$(a + 2hq + bq^2)\frac{\partial^2 u}{\partial \xi^2} + 2[ar + bsq + h(rq + s)]\frac{\partial^2 u}{\partial \xi \partial \eta} +$$
$$+ (ar^2 + bs^2 + 2hrs)\frac{\partial^2 u}{\partial \eta^2} = 0. \tag{45}$$

If q is now chosen to be the root of

$$a + bq^2 + 2hq = 0 \tag{46}$$

then, since $ab - h^2 = 0$ by assumption,

$$q = -\frac{h}{b} \quad \text{(twice)}. \tag{47}$$

Hence in virtue of (46) the first term of (45) is zero. Likewise the second term of (45) is zero since

$$ar + bsq + h(rq + s) = (ab - h^2)\frac{r}{b} = 0. \tag{48}$$

Consequently, provided r and s are not both zero, (45) becomes

$$\frac{\partial^2 u}{\partial \eta^2} = 0. \tag{49}$$

By direct integration

$$u = F(\xi) + \eta G(\xi), \tag{50}$$

where F and G are arbitrary functions. Hence, since $p = 1$ and $q = -h/b \ (= \lambda, \text{ say})$ we have from (12)

$$\xi = x + \lambda y,$$
$$\eta = rx + sy, \tag{51}$$

where r and s are arbitrary (but not both zero). Choosing $r=0$, $s=1$ for simplicity, we have

$$u = F(x + \lambda y) + yG(x + \lambda y) \tag{52}$$

as the general solution of (11) in the parabolic case, where λ is the double root of

$$a + 2h\lambda + b\lambda^2 = 0. \tag{53}$$

Example 5. The equation

$$\frac{\partial^2 u}{\partial x^2} + 4 \frac{\partial^2 u}{\partial x \partial y} + 4 \frac{\partial^2 u}{\partial y^2} = 0 \tag{54}$$

is parabolic since $a=1$, $h=2$, $b=4$ and $ab - h^2 = 0$. Equation (53) becomes

$$1 + 4\lambda + 4\lambda^2 = 0, \tag{55}$$

whence

$$\lambda = -\tfrac{1}{2} \quad \text{(twice)}.$$

Hence by (52) the general solution of (54) is

$$u = F(x - \tfrac{1}{2}y) + yG(x - \tfrac{1}{2}y). \tag{56}$$

In conclusion we remark that inhomogeneous equations of the type

$$a \frac{\partial^2 u}{\partial x^2} + 2h \frac{\partial^2 u}{\partial x \partial y} + b \frac{\partial^2 u}{\partial y^2} = f(x, y), \tag{57}$$

where a, h, b are constants and $f(x, y)$ is a given function, may be solved by an extension of the D-operator method used for ordinary differential equations. However, in practice this method is not particularly useful and we shall not develop it here. Overall the method of obtaining the general solutions as shown in this section is not particularly vital since (as discussed in 1.1) the general solutions are usually far too difficult to fit to the boundary conditions.

2.3 Boundary conditions

In the preceding sections we have concentrated on particular types of equations and their general solutions, remarking that such solutions are often of little value in boundary-value problems owing to the difficulty of determining the unknown functions. We now discuss boundary conditions and solutions in more detail with special reference to physical problems.

The mathematical representation of a physical phenomenon by a partial differential equation and a set of boundary conditions is said to be well-posed or well-formulated provided two criteria are satisfied. Firstly, the solution should be unique, since our experience of nature is such that a given set of circumstances leads to just one outcome. Secondly, the solution obtained should be stable. In other words, a small change in the given boundary conditions should produce only a correspondingly small change in the solution. This is vital since, when the boundary conditions are arrived at by experiment, certain small observational errors in their values will always exist and these errors should not lead to large changes in the solution. (A similar situation arises with sets of linear algebraic equations, where under certain circumstances the equations may be ill-conditioned – that is, small changes in the coefficients may produce violent changes in the solutions).† Consider, for example, Laplace's equation in two dimensions

$$\frac{\partial^2 u}{\partial x^2} + \frac{\partial^2 u}{\partial y^2} = 0. \tag{58}$$

We now search for a solution which satisfies the boundary conditions

$$u(x, 0) = \frac{\sin nx}{n}, \quad \left(\frac{\partial u}{\partial y}\right)_{y=0} = 0, \tag{59}$$

where n is some parameter.

Such a solution is easily verified to be

$$u(x, y) = \frac{1}{n} \sin nx \cosh ny. \tag{60}$$

However, as $n \to \infty$ the boundary conditions converge to

$$u(x, 0) = 0, \quad \left(\frac{\partial u}{\partial y}\right)_{y=0} = 0, \tag{61}$$

and these, together with (58), imply, by Taylor's series, $u(x, y) = 0$. However, as $n \to \infty$, $u(x, y)$ as given by (60) becomes infinitely large. The problem defined by (58) and (59) is not, therefore, well-posed and could not be associated with a physical phenomenon.

† See, for example, the author's *Introduction to Matrices, Sets and Groups for Science Students*, Longmans 1965 (Chapter 4).

Much work has been carried out to determine the types of boundary conditions which, when imposed on linear partial differential equations, lead to unique and stable solutions. Such an analysis is too difficult and lengthy to be given here, but a full account may be found in reference [13] in the list of Further Reading at the end of this book.

There are three main types of boundary conditions which arise frequently in the description of physical phenomena. These are

(a) Dirichlet conditions, where u is specified at each point of a boundary of a region (for example, the bounding curve of a plane region, or the surface of a three-dimensional domain). The problem of solving Laplace's equation $\nabla^2 u = 0$ inside a region with prescribed values of u on the boundary is called the Dirichlet problem.

(b) Neumann conditions, where values of the normal derivative $\frac{du}{dn}$ of the function are prescribed on the boundary.

(c) Cauchy conditions. Here, if one of the independent variables is t (time, say), and the values of both u and $\frac{\partial u}{\partial t}$ on a boundary $t = 0$ $\left(\text{that is, the } \textit{initial} \text{ values of } u \text{ and } \frac{\partial u}{\partial t}\right)$ are given, then the boundary conditions are of Cauchy type with respect to the variable t.

In the case of the one-dimensional wave equation representing, say, the transverse oscillations of a stretched string (see Chapter 1, 1.2), the Cauchy conditions correspond to giving the initial values of both the transverse displacement u and the transverse velocity $\frac{\partial u}{\partial t}$ of the string. These conditions can be shown to be necessary and sufficient for the existence of an unique solution (see 2.5).

2.4 Laplace's equation and the Dirichlet problem

Whereas it is difficult, if not impossible, to prove uniqueness theorems for most systems of partial differential equations and boundary conditions, in some special cases it is relatively easy. The Laplace equation is of such significance in both mathematical physics and in pure mathematics (arising as it does in complex variable theory, potential theory and harmonic vector theory, to name but a few instances) that we devote this section to proving certain results about its solutions. These results can then be used to demonstrate the uniqueness of the solution of the Dirichlet problem (see 2.3 (a)).

For simplicity we restrict the analysis to the case of two independent variables (x, y), the case of three independent variables being essentially similar.

Any function $u(x, y)$ which satisfies Laplace's equation $\nabla^2 u = 0$ is called a *harmonic* function. In the one-dimensional (i.e. one independent variable) case, $\nabla^2 u$ becomes $\dfrac{d^2 u}{dx^2}$ and consequently one-dimensional harmonic functions satisfy

$$\frac{d^2 u}{dx^2} = 0. \tag{62}$$

Solutions of (62) which are continuous are the linear functions $(u = Ax + B)$ and are uniquely determined by the values at the two end points of some interval of x. In the one-dimensional case these end points constitute the boundary of the region (the interval of x) and consequently the solution of the one-dimensional Laplace equation is uniquely determined by prescribing values of u on the boundary. The solution of the one-dimensional Dirichlet problem is therefore unique.

In two dimensions we first prove that the maximum and minimum values of a function $u(x, y)$ which is harmonic in a region R are attained in the boundary of R. Consider a function $v(x, y)$ such that within R

$$\nabla^2 v = \frac{\partial^2 v}{\partial x^2} + \frac{\partial^2 v}{\partial y^2} > 0. \tag{63}$$

If $v(x, y)$ had a maximum in R we would have at that maximum point $\dfrac{\partial^2 v}{\partial x^2} \leqslant 0$, $\dfrac{\partial^2 v}{\partial y^2} \leqslant 0$ simultaneously, which would give

$$\nabla^2 v \leqslant 0. \tag{64}$$

This is in contradiction to the assumption (63); therefore the maximum value of v occurs on the boundary of R. Similarly by considering the Laplacian of the negative of v we can show that the minimum value of v occurs on the boundary R.

If now $u(x, y)$ is a harmonic function its Laplacian $\nabla^2 u$ is zero. Clearly $u(x, y)$ may be changed by an arbitrarily small amount in such a way that its Laplacian will become positive. Consider, for example, the function

$$v = u + \varepsilon(x^2 + y^2), \tag{65}$$

where ε is an arbitrarily small positive quantity. Then

$$\nabla^2 v = 4\varepsilon > 0. \tag{66}$$

Now the addition of a sufficiently small positive quantity cannot change the property of a function having a maximum within a region R. If then a harmonic function $u(x, y)$ were to have a maximum inside R then by adding $\varepsilon(x^2 + y^2)$ as in (65) we would obtain a function with a positive Laplacian (66). But as already shown such a function cannot have a maximum inside R. Hence a harmonic function in R cannot possess a maximum inside R.

Similar arguments show that a harmonic function in R cannot have a minimum inside R.

Consequently the maximum and minimum values of u are obtained on the boundary of R.

Using this result we can now prove the uniqueness of the solution of the Dirichlet problem. For suppose $u(x, y)$ is given on the boundary of R. If $u_1(x, y)$ and $u_2(x, y)$ are two supposedly different harmonic functions (that is – solutions of $\nabla^2 u = 0$), then their difference is also a harmonic function which takes zero values on the boundary. Since the maximum and minimum values of all functions harmonic within R must occur on the boundary of R, we have

$$u_1(x, y) - u_2(x, y) = 0 \tag{67}$$

inside R. This shows that there is in fact only one solution which is completely determined by the values of $u(x, y)$ on the boundary.

Similar arguments can be applied to the heat conduction equation. An alternative method of proving the uniqueness of solutions of the Laplace equation, the heat conduction equation and the wave equation depends on the use of Green's theorem, but we will not discuss this approach here (see, however, reference [3]).

As a rule, unique and stable solutions of both elliptic and parabolic equations may be obtained by imposing either Dirichlet or Neumann boundary conditions, whereas unique stable solutions of hyperbolic equations arise by imposing Cauchy type conditions.

24

Again we refer the reader to reference [13] in the list of Further Reading for a full discussion.

2.5 D'Alembert's solution of the wave equation

We now solve the one-dimensional wave equation

$$\frac{\partial^2 u}{\partial x^2} = \frac{1}{c^2}\frac{\partial^2 u}{\partial t^2}, \qquad (68)$$

where $u = u(x, t)$, subject to the Cauchy initial conditions

$$u(x, 0) = f(x), \qquad (69)$$

$$\left(\frac{\partial u}{\partial t}\right)_{t=0} = g(x). \qquad (70)$$

The general solution of (68) is, by (31),

$$u(x, t) = F(x + ct) + G(x - ct), \qquad (71)$$

where F and G are arbitrary twice differentiable functions of their arguments.

Hence, using (69), (70) and (71), we have

$$F(x) + G(x) = f(x), \qquad (72)$$

$$cF'(x) - cG'(x) = g(x), \qquad (73)$$

where the primes refer to derivatives.

Integrating this last equation we have

$$F(x) - G(x) = \frac{1}{c}\int_a^x g(\xi)d\xi, \qquad (74)$$

where the constant of integration has been incorporated in the lower limit by introducing an arbitrary constant a. From (72) and (74) we obtain

$$F(x) = \tfrac{1}{2}f(x) + \frac{1}{2c}\int_a^x g(\xi)d\xi, \qquad (75)$$

and

$$G(x) = \tfrac{1}{2}f(x) - \frac{1}{2c}\int_a^x g(\xi)d\xi. \qquad (76)$$

Hence

$$F(x + ct) = \tfrac{1}{2}f(x + ct) + \frac{1}{2c}\int_a^{x+ct} g(\xi)d\xi, \qquad (77)$$

and

$$G(x - ct) = \tfrac{1}{2}f(x - ct) - \frac{1}{2c}\int_a^{x-ct} g(\xi)d\xi. \qquad (78)$$

Finally, therefore, the solution of (68) subject to (69) and (70) is

$$u(x, t) = \tfrac{1}{2}[f(x + ct) + f(x - ct)] + \frac{1}{2c} \int_{x-ct}^{x+ct} g(\xi)d\xi. \qquad (79)$$

This solution is called D'Alembert's solution. We note that the value of $u(x, t)$ depends only on the initial values at points between $x - ct$ and $x + ct$ and not at all on initial values outside this range. This range, or interval, is called the domain of dependence of the variables (x, t). Clearly the solution (79) is unique, and moreover is stable since it depends continuously on the values of the initial conditions.

PROBLEMS 2

1. Determine the nature of each of the following equations (i.e. whether elliptic, parabolic or hyperbolic) and obtain the general solution in each case:

   (a) $3 \dfrac{\partial^2 u}{\partial x^2} + 4 \dfrac{\partial^2 u}{\partial x \partial y} - \dfrac{\partial^2 u}{\partial y^2} = 0$,   (b) $\dfrac{\partial^2 u}{\partial x^2} - 2 \dfrac{\partial^2 u}{\partial x \partial y} + \dfrac{\partial^2 u}{\partial y^2} = 0$,

   (c) $4 \dfrac{\partial^2 u}{\partial x^2} + \dfrac{\partial^2 u}{\partial y^2} = 0$,   (d) $\dfrac{\partial^2 u}{\partial x^2} + 4 \dfrac{\partial^2 u}{\partial x \partial y} + 4 \dfrac{\partial^2 u}{\partial y^2} = 0$,

   (e) $\dfrac{\partial^2 u}{\partial y^2} + 2 \dfrac{\partial^2 u}{\partial x^2} = 0$.

2. A function $u(r, t)$ satisfies the equation

$$\frac{1}{r^2} \frac{\partial}{\partial r}\left(r^2 \frac{\partial u}{\partial r}\right) = \frac{1}{c^2} \frac{\partial^2 u}{\partial t^2},$$

where c is a constant. By introducing the new dependent variable $v(r, t) = ru(r, t)$, and writing $\xi = r + ct$, $\eta = r - ct$, reduce this equation to

$$\frac{\partial^2 v}{\partial \xi \partial \eta} = 0.$$

Hence show that the general solution $u(r, t)$ has the form

$$u(r, t) = \frac{1}{r}[f(r + ct) + g(r - ct)],$$

where f and g are arbitrary (twice differentiable) functions.

26

3. Solve the following boundary value problems by first obtaining the general solutions of the partial differential equations.

(a) $\dfrac{\partial^2 u}{\partial x^2} = \dfrac{1}{c^2}\dfrac{\partial^2 u}{\partial t^2}$, given $u(x, 0) = 0$, $\left(\dfrac{\partial u}{\partial t}\right)_{t=0} = \dfrac{1}{1+x^2}$.

(b) $\dfrac{\partial^2 u}{\partial x^2} = 2xy$, given $u(0, y) = y^2$, and $\left(\dfrac{\partial u}{\partial x}\right)_{x=0} = y$.

(c) $\dfrac{\partial^2 u}{\partial x \partial y} = 1$, given $u = 0$, $\dfrac{\partial u}{\partial x} = 0$ on $x + y = 0$.

CHAPTER 3

Orthonormal Functions

3.1 Superposition of solutions

In the preceding two chapters we have discussed from a somewhat physical standpoint the origin and nature of certain types of equations and boundary conditions. From now on we shall be primarily concerned with the methods of solution of boundary value problems. As remarked in the previous chapter, the general solutions of partial differential equations are of little use owing to the difficulty of choosing the arbitrary functions in such a way that the boundary conditions are satisfied. This difficulty can be avoided for some linear partial differential equations by various different techniques, one of which is based on the Principle of Superposition which states that if each of the n functions u_1, u_2, $\ldots$, u_n satisfies a linear homogeneous partial differential equation, then an arbitrary linear combination of these functions

$$u = c_1 u_1 + c_2 u_2 + \ldots + c_n u_n = \sum_{i=1}^{n} c_i u_i, \qquad (1)$$

where c_i are constants, also satisfies the equation. This principle is readily verified in the following way: suppose u_1 and u_2 are any two functions of a set of functions and that L is some operator for which

$$L(c_1 u_1 + c_2 u_2) = c_1 L u_1 + c_2 L u_2, \qquad (2)$$

where c_1 and c_2 are constants. Such an operator is called a linear operator. Clearly if u_3 is another function of the set then

$$
\begin{aligned}
L(c_1 u_1 + c_2 u_2 + c_3 u_3) &= L(c_1 u_1 + c_2 u_2) + L(c_3 u_3) \\
&= L(c_1 u_1 + c_2 u_2) + c_3 L u_3 \quad \text{(by (2))} \\
&= c_1 L u_1 + c_2 L u_2 + c_3 L u_3, \quad \text{(by (2))}.
\end{aligned}
\qquad (3)
$$

In a like manner we may readily show that for an arbitrary linear combination of n functions u_1, u_2, $\ldots$, u_n

$$L\left(\sum_{i=1}^{n} c_i u_i \right) = \sum_{i=1}^{n} c_i L u_i. \qquad (4)$$

28

The D-operator $\left(D \equiv \dfrac{d}{dx}\right)$ is a linear differential operator on the set of all functions of one variable which are at least once differentiable. Similarly $\dfrac{\partial}{\partial x}, \dfrac{\partial}{\partial y}$ are linear differential operators on the set of all functions of two independent variables (x, y) which are at least once differentiable with respect to both x and y.

We now see that every linear homogeneous partial differential equation has the form

$$Lu = 0, \tag{5}$$

where L is a linear operator and u is the dependent variable. For example, the reader may easily verify that $\dfrac{\partial^2}{\partial x^2}$ and $\dfrac{\partial^2}{\partial y^2}$ are both linear operators, and that Laplace's equation in two-dimensions may be written in the form (5) where L is the linear differential operator

$$\frac{\partial^2}{\partial x^2} + \frac{\partial^2}{\partial y^2} \quad (= \nabla^2). \tag{6}$$

Likewise, the general linear homogeneous partial differential equation of the second order

$$a\,\frac{\partial^2 u}{\partial x^2} + 2h\,\frac{\partial^2 u}{\partial x \partial y} + b\,\frac{\partial^2 u}{\partial y^2} + 2f\,\frac{\partial u}{\partial x} + 2g\,\frac{\partial u}{\partial y} + eu = 0 \tag{7}$$

(see Chapter 2), where a, h, b, f, g and e may be functions of x and y, may be written in the form (5) where now

$$L \equiv a\,\frac{\partial^2}{\partial x^2} + 2h\,\frac{\partial^2}{\partial x \partial y} + b\,\frac{\partial^2}{\partial y^2} + 2f\,\frac{\partial}{\partial x} + 2g\,\frac{\partial}{\partial y} + e. \tag{8}$$

We now see from (4) and (5) that if u_i $(i = 1, 2, 3, \ldots, n)$ satisfy the partial differential equation $Lu_i = 0$, then an arbitrary linear combination of such solutions is also a solution. This is the Principle of Superposition.

Now if it is possible to find such a set of solutions u_i for any given linear homogeneous partial differential equation it may be possible by choosing the appropriate linear combination to satisfy all the given boundary conditions. In order to generate such a set of solutions we use a method known as the method of separation of variables. Here the dependent variable is assumed to be a product of

29

functions, each of which depends on just one of the independent variables. The combination of the method of separation of variables and the superposition of solutions is usually known as Fourier's method, and we now illustrate the general nature of this method by the following example, leaving aside the justification of certain points until later sections of this chapter.

Example 1. To obtain the solution of the one-dimensional wave equation

$$\frac{\partial^2 u}{\partial x^2} = \frac{1}{c^2} \frac{\partial^2 u}{\partial t^2} \tag{9}$$

which satisfies the Cauchy boundary conditions

$$u(0, t) = u(l, t) = 0, \quad t \geqslant 0, \tag{10}$$

$$u(x, 0) = f(x), \quad 0 \leqslant x \leqslant l, \tag{11}$$

$$\left(\frac{\partial u}{\partial t}\right)_{t=0} = g(x), \quad 0 \leqslant x \leqslant l, \tag{12}$$

where f and g are given functions, and l is a given constant.

We now assume a separable solution of (9) of the form

$$u(x, t) = X(x)T(t), \tag{13}$$

where X is a function of x only, and T is a function of t only. In this way (9) becomes

$$\frac{1}{X} \frac{d^2 X}{dx^2} = \frac{1}{c^2 T} \frac{d^2 T}{dt^2}. \tag{14}$$

Now the left-hand side of this equation is a function of x only and the right-hand side is independent of x. Hence it follows that (14) can only be true if both left- and right-hand sides have the same constant value. We write then

$$\frac{1}{X} \frac{d^2 X}{dx^2} = \frac{1}{c^2 T} \frac{d^2 T}{dt^2} = k, \tag{15}$$

where k is an arbitrary constant. In this way the pair of ordinary differential equations

$$\frac{d^2 X}{dx^2} = kX, \tag{16}$$

$$\frac{d^2 T}{dt^2} = kc^2 T \tag{17}$$

30

are obtained. The functions X and T may now be found by solving these equations, but we must at the same time ensure that the resulting solution $u(x, t)$ given by (13) satisfies the boundary values. By (10) and (13)

$$u(0, t) = X(0)T(t) = 0, \quad \text{for all } t.$$
$$u(l, t) = X(l)T(t) = 0, \quad \text{for all } t. \tag{18}$$

Hence, taking $T(t) \not\equiv 0$ (thus excluding the trivial solution $u(x, t) \equiv 0$), we have

$$X(0) = X(l) = 0. \tag{19}$$

Three possible cases now arise, namely k positive, negative and zero.

$k = 0$. In this case (16) gives

$$X(x) = Ax + B, \tag{20}$$

whence, using (19), we find $A = B = 0$. Consequently $X(x) \equiv 0$ and hence $u(x, t) = 0$.

k positive $(= \omega^2)$. Here (16) leads to

$$X(x) = Ae^{\omega x} + Be^{-\omega x}, \tag{21}$$

which together with (19) leads to $A = B = 0$. Again this results in the trivial solution $u(x, t) = 0$.

k negative $(= -\omega^2)$. In this case (16) gives

$$X(x) = A \cos \omega x + B \sin \omega x \tag{22}$$

from which we may derive, using (19), the non-trivial solution

$$X(x) = B \sin \omega x, \quad (B \text{ arbitrary}), \tag{23}$$

with

$$\sin \omega l = 0. \tag{24}$$

Equation (24) shows that the parameter ω has the form

$$\omega = \frac{r\pi}{l}, \quad r = 1, 2, 3 \ldots \tag{25}$$

(the case $r = 0$, which gives $\omega = 0$, being excluded since this would again result in the trivial solution $u(x, t) = 0$).

Now solving (17) with $k = -\omega^2$ we have

$$T(t) = C \cos \omega ct + D \sin \omega ct, \tag{26}$$

where C and D are integration constants. Hence the solution (13) takes the form, using (23) and (26),

$$u(x, t) = X(x)T(t) = \sin \omega x(C \cos \omega ct + D \sin \omega ct), \tag{27}$$

the arbitrary constant B having been put equal to unity for simplicity.

However, we see from (25) that there exists an *infinite* set of discrete

31

values of ω (the eigenvalues or characteristic values), and consequently to each value of ω there will correspond a particular solution (the eigenfunction or characteristic function) having the form (27). These are

$$
\left.
\begin{aligned}
\omega_1 &= \frac{\pi}{l}, \; u_1(x,\, t) = \sin \frac{\pi x}{l} \left(C_1 \cos \frac{\pi c t}{l} + D_1 \sin \frac{\pi c t}{l} \right), \\[4pt]
\omega_2 &= \frac{2\pi}{l}, \; u_2(x,\, t) = \sin \frac{2\pi x}{l} \left(C_2 \cos \frac{2\pi c t}{l} + D_2 \sin \frac{2\pi c t}{l} \right), \\
&\quad\vdots \\
\omega_r &= \frac{r\pi}{l}, \; u_r(x,\, t) = \sin \frac{r\pi x}{l} \left(C_r \cos \frac{r\pi c t}{l} + D_r \sin \frac{r\pi c t}{l} \right), \\
&\quad\vdots
\end{aligned}
\right\} \tag{28}
$$

and so on, $C_1, C_2, \ldots, C_r, \ldots, D_1, D_2, \ldots, D_r, \ldots$ being arbitrary constants. Each of these expressions for $u(x, t)$ is a solution of the wave equation (9) satisfying the boundary conditions (10). Now since (9) is a linear equation any linear combination of such solutions is also a solution. Accordingly we take the linear combination

$$
u(x, t) = \sum_{r=1}^{\infty} \left(C_r \cos \frac{r\pi c t}{l} + D_r \sin \frac{r\pi c t}{l} \right) \sin \frac{r\pi x}{l} \tag{29}
$$

as the general solution of (9) satisfying the boundary conditions (10). The arbitrary constants C_r and D_r in this solution must now be chosen so that the boundary conditions at $t = 0$ (i.e. (11) and (12)) are satisfied.

Consider first (11) which requires

$$
u(x, 0) = f(x), \quad 0 \leqslant x \leqslant l. \tag{30}
$$

Then putting $t = 0$ in (29) we have

$$
f(x) = \sum_{r=1}^{\infty} C_r \sin \frac{r\pi x}{l}. \tag{31}
$$

Similarly (12), which requires

$$
\left(\frac{\partial u}{\partial t} \right)_{t=0} = g(x), \quad 0 \leqslant x \leqslant l, \tag{32}
$$

is satisfied by differentiating (29) with respect to t and then putting $t = 0$. In this way

$$
g(x) = \frac{\pi c}{l} \cdot \sum_{r=1}^{\infty} r D_r \sin \frac{r\pi x}{l}. \tag{33}
$$

The sets of coefficients C_r and D_r may now be determined from (31) and (33) by a Fourier series technique. Consequently (see next section)

$$C_r = \frac{2}{l} \int_0^l f(x) \sin \frac{r\pi x}{l} \, dx \qquad (34)$$

and

$$D_r = \frac{2}{r\pi c} \int_0^l g(x) \sin \frac{r\pi x}{l} \, dx \qquad (35)$$

where $r = 1, 2, 3, \ldots$

Hence, finally, substituting (34) and (35) into (29) we have the solution

$$u(x, t) = \sum_{r=1}^{\infty} \left\{ \left[\frac{2}{l} \int_0^l f(x') \sin \frac{r\pi x'}{l} \, dx' \right] \cos \frac{r\pi ct}{l} \sin \frac{r\pi x}{l} + \left[\frac{2}{r\pi c} \int_0^l g(x') \sin \frac{r\pi x'}{l} \, dx' \right] \sin \frac{r\pi ct}{l} \sin \frac{r\pi x}{l} \right\}, \qquad (36)$$

where we have written x' for the variable of integration to distinguish it from the independent variable x. This function is a solution of (9) satisfying the boundary conditions (10), (11) and (12).

3.2 Orthonormal functions

The fact that (34) and (35) may be written down is a direct consequence of (31) and (33) being expressed as sums of sines. For since by direct integration

$$\int_0^l \sin \frac{r\pi x}{l} \sin \frac{s\pi x}{l} \, dx = \begin{cases} 0, & r \neq s, \\ l/2, & r = s \, (\neq 0), \end{cases} \qquad (37)$$

where r and s are positive integers, we only have to multiply each side of the series (31) and (33) by $\sin \frac{s\pi x}{l}$ and integrate term-by-term from 0 to l to obtain (34) and (35). This is the method of Fourier series and assumes that the functions so expanded satisfy certain conditions (the Dirichlet conditions) which ensure that the series converge to the given functions at each point (except at points of discontinuity).†

† See, for example, the author's *Mathematical Methods for Science Students*, 2nd Edition Longman 1973 (Chapter 15).

Now (37) may be written as

$$\int_0^l \left(\sqrt{\left(\frac{2}{l}\right)}\sin\frac{r\pi x}{l}\right)\left(\sqrt{\left(\frac{2}{l}\right)}\sin\frac{s\pi x}{l}\right)dx = \begin{cases} 0, & r\neq s, \\ 1, & r=s\ (\neq 0), \end{cases} \qquad (38)$$

whence we see that if we write

$$\varphi_r(x) = \sqrt{\left(\frac{2}{l}\right)}\sin\frac{r\pi x}{l}, \quad r=1, 2, 3,\ldots, \qquad (39)$$

then

$$\int_0^l \varphi_r(x)\varphi_s(x)dx = \begin{cases} 0, & r\neq s, \\ 1, & r=s\ (\neq 0). \end{cases} \qquad (40)$$

Functions $\varphi_r(x)$ for which the integral of $\varphi_r(x)\varphi_s(x)$ for $r\neq s$ vanishes over the interval $(0, l)$ are said to be orthogonal functions on the interval $(0, l)$. Such functions are also said to be normalised to unity on $(0, l)$ if for $r=s\ (\neq 0)$ the same integral is equal to unity. These are the two instances in (40). Functions (such as (39), for example) which are orthogonal and normalised to unity on $(0, l)$ are said to be orthonormal on $(0, l)$. Consequently any set of functions $\varphi_r(x)$ $(r=1, 2, 3\ldots)$ is an orthonormal set of functions on $(0, l)$ if

$$\int_0^l \varphi_r(x)\varphi_s(x)dx = \delta_{rs}, \quad (r, s=1, 2, 3,\ldots) \qquad (41)$$

where δ_{rs} is the Kronecker delta symbol defined by

$$\delta_{rs} = \begin{cases} 0, & r\neq s, \\ 1, & r=s. \end{cases} \qquad (42)$$

In general a set of functions $\varphi_r(x)$ is said to be orthonormal on the interval (a, b) with respect to a weight function $\omega(x)$ if

$$\int_a^b \omega(x)\varphi_r(x)\varphi_s(x)dx = \delta_{rs}. \quad (r, s=1, 2, 3,\ldots). \qquad (43)$$

For example, the infinite set of functions

$$\frac{1}{\sqrt{(2\pi)}}, \quad \frac{\cos x}{\sqrt{\pi}}, \quad \frac{\sin x}{\sqrt{\pi}},\ldots, \quad \frac{\cos nx}{\sqrt{\pi}}, \quad \frac{\sin nx}{\sqrt{\pi}},\ldots \qquad (44)$$

is orthonormal on the interval $(-\pi, \pi)$ with respect to a weight function $\omega(x) = 1$. In later chapters of this book we will meet other sets of orthonormal functions, one of which is the sequence of Legendre polynomials (see Chapter 5) defined by

$$P_n(x) = \frac{1}{2^n n!} \frac{d^n}{dx^n}[(x^2-1)^n], \qquad (45)$$

where $n = 0, 1, 2, \ldots$. The first few polynomials of this sequence are

$$P_0(x) = 1, \quad P_1(x) = x, \quad P_2(x) = \tfrac{1}{2}(3x^2 - 1), \quad P_3(x) = \tfrac{1}{2}(5x^3 - 3x). \quad (46)$$

It may be verified that the set of polynomials

$$\sqrt{\left(\frac{2n+1}{2}\right)} P_n(x), \quad n = 0, 1, 2\ldots, \quad (47)$$

are orthonormal on the interval $(-1, 1)$ with respect to the weight function $\omega(x) = 1$ since it may be shown that

$$\int_{-1}^{1} \sqrt{\left(\frac{2n+1}{2}\right)} P_n(x) \sqrt{\left(\frac{2m+1}{2}\right)} P_m(x) dx = \delta_{nm}.$$
$$(n, m = 0, 1, 2, \ldots) \quad (48)$$

Instances of weight functions which are not constants (unlike the previous two examples) arise with orthogonal sets of Bessel functions (see, for example, Chapter 5), and with the Tchebysheff polynomials

$$T_n(x) = \cos (n \cos^{-1} x), \quad (n = 0, 1, 2, \ldots), \quad (49)$$

which are orthogonal on the interval $(-1, 1)$ with respect to the weight function

$$\omega(x) = \frac{1}{\sqrt{(1-x^2)}}. \quad (50)$$

3.3 Expansion of a function in a series of orthonormal functions

Now just as in three-dimensional space where every vector $\mathbf{A}$ (say) may be represented by a linear combination of three orthogonal vectors $\mathbf{e}_1, \mathbf{e}_2, \mathbf{e}_3$, each of unit length, so that

$$\mathbf{A} = a_1\mathbf{e}_1 + a_2\mathbf{e}_2 + a_3\mathbf{e}_3, \quad (51)$$

where a_1, a_2 and a_3 are constants (see Fig. 3.1) so it may be possible to represent an arbitrary function $f(x)$ on an interval (a, b) as a linear combination of an *infinite* number of orthonormal functions $\varphi_r(x)$ such that

$$f(x) = c_1\varphi_1(x) + c_2\varphi_2(x) + \ldots = \sum_{r=1}^{\infty} c_r\varphi_r(x), \quad (52)$$

where the c_r are constants. This type of expansion is often referred to as a generalised Fourier series. Assuming (as in (43)) that the $\varphi_r(x)$ are orthonormal on an interval (a, b) with respect to a weight function $\omega(x)$ so that

$$\int_a^b \omega(x)\varphi_r(x)\varphi_s(x)dx = \delta_{rs}, \quad (53)$$

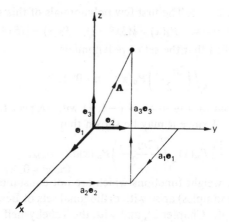

Fig. 3.1

then, using (52), we have

$$\omega(x)f(x)\varphi_r(x) = c_1\omega(x)\varphi_1(x)\varphi_r(x) + c_2\omega(x)\varphi_2(x)\varphi_r(x) + \ldots +$$
$$+ c_r\omega(x)\varphi_r(x)\varphi_r(x) + c_{r+1}\omega(x)\varphi_{r+1}(x)\varphi_r(x) + \ldots \quad (54)$$

Hence integrating each side of (54) from a to b we find

$$\int_a^b \omega(x)f(x)\varphi_r(x)dx = c_1\int_a^b \omega(x)\varphi_1(x)\varphi_r(x)dx +$$
$$+ c_2\int_a^b \omega(x)\varphi_2(x)\varphi_r(x)dx + \ldots +$$
$$+ c_r\int_a^b \omega(x)\varphi_r(x)\varphi_r(x)dx + \ldots \quad (55)$$

which, using (53), gives

$$\int_a^b \omega(x)f(x)\varphi_r(x)dx = c_r. \quad (56)$$

In determining the coefficients c_r in this way we have assumed that the expansion of $f(x)$ in the form (52) is valid. Strictly we should not have an equality sign in (52) since at this stage it is not known whether the series will converge, and, if it does, whether it will converge to $f(x)$. For such an expansion to be valid the orthonormal set of functions must satisfy the condition of *completeness*.

36

An orthonormal set of functions $\varphi_r(x)$ is said to be complete if it is impossible to add to it one other *non-zero* function which is orthogonal to each of the $\varphi_r(x)$. By analogy with the representation of a vector in three-dimensional space (see Fig. 3.1) we see that it is impossible to add any other vector to the set e_1, e_2 and e_3 which is orthogonal to each. In other words every non-zero vector A may be represented in terms of the complete set of vectors e_1, e_2, e_3. However, the set of two orthogonal unit vectors e_1, e_2 would not form a complete set in three-dimensions since it is possible to find another vector, e_3, which is orthogonal to both e_1 and e_2. For an incomplete set such as e_1, e_2, any vector not in the plane of e_1 and e_2 cannot be represented by a linear combination of these two vectors.

We can readily give an example of an incomplete orthogonal set of functions by removing any one member of the set (44) – say, the function $\dfrac{\sin x}{\sqrt{\pi}}$. The remaining infinite system of functions

$$\frac{1}{\sqrt{(2\pi)}}, \quad \frac{\cos x}{\sqrt{\pi}}, \quad \frac{\cos 2x}{\sqrt{\pi}}, \quad \frac{\sin 2x}{\sqrt{\pi}}, \cdots, \tag{57}$$

is still an orthonormal system but is incomplete since we can now add to the set the function $\dfrac{\sin x}{\sqrt{\pi}}$ which is orthogonal to each member of (57).

The whole problem of proving completeness of orthonormal sets of functions and the related question of convergence of series of the type (52) leads into a branch of mathematics called functional analysis in which such concepts as Hilbert space and Lesbesgue integration are of importance. We shall not attempt to discuss these ideas here, but simply remark that all the sets of orthonormal functions which commonly arise in mathematical physics, and certainly all those met in this book, have been proved to be complete. A full treatment of the convergence of series of orthonormal functions is beyond our present scope, but, in general, functions satisfying the Dirichlet conditions may be expanded in such series and under these conditions the series will converge to the function at each point of the interval (a, b) where the function is continuous. At any point within (a, b) where the function has a finite discontinuity the series will converge to the mean of the left- and right-hand limits.

In some problems it is often necessary to expand an arbitrary function of two variables, $f(x, y)$, in a series of two sets of real orthonormal functions $\varphi_r(x)$, $\theta_r(y)$ $(r = 1, 2, 3, \ldots)$ such that

$$f(x, y) = \sum_{r=1}^{\infty} \sum_{s=1}^{\infty} c_{rs}\varphi_r(x)\theta_s(y), \tag{58}$$

where c_{rs} are constants, and where $\varphi_r(x)$ is an orthonormal set on the interval $a \leqslant x \leqslant b$ with respect to the weight function $\omega(x)$, such that

$$\int_a^b \omega(x)\varphi_r(x)\varphi_s(x)dx = \delta_{rs}, \quad (r, s = 1, 2, 3, \ldots), \tag{59}$$

and where $\theta_r(y)$ is an orthonormal set on the interval $c \leqslant y \leqslant d$, with respect to the weight function $v(y)$ such that

$$\int_c^d v(y)\theta_r(y)\theta_s(y)dy = \delta_{rs}. \quad (r, s = 1, 2, 3, \ldots). \tag{60}$$

Proceeding in a similar way to the case of one orthonormal set dealt with in the first part of this section, it is easily found that formally the coefficients c_{rs} are given by

$$c_{rs} = \int_c^d v(y)\theta_s(y)\left[\int_a^b f(x, y)\omega(x)\varphi_r(x)dx\right]dy. \tag{61}$$

The conditions under which the double series (58) with coefficients c_{rs} given by (61) is convergent to the function $f(x, y)$ will not be discussed here, but in general most functions met in mathematical physics are sufficiently well-behaved to be expanded in this way.

3.4 The Sturm-Liouville equation

In Example 1 of 3.1, we saw that the technique of separation of variables led to two ordinary differential equations for $X(x)$ and $T(t)$, and that in solving equation (16) for $X(x)$ subject to the given boundary conditions a set of orthonormal functions arose. The method of separation of variables when applied to second-order linear partial differential equations frequently leads to an ordinary differential equation of the type

$$\frac{d}{dx}\left(p\frac{d\varphi(x)}{dx}\right) + (q + \lambda r)\varphi(x) = 0, \tag{62}$$

where p, q and r are given functions of the independent variable x in an interval $a \leqslant x \leqslant b$ (say), λ is a parameter, and $\varphi(x)$ is the dependent variable. This equation is known as the Sturm-Liouville equation and generates, when certain types of boundary conditions are applied, various sets of orthogonal functions which frequently arise in the solution of partial differential equations. (Equation (16) of this chapter is a special case of (62) obtained by putting $p = 1$, $q = 0$, $r = 1$, $\lambda = -k$, and letting $\varphi(x) \equiv X(x)$.) It will turn out that in general (62), when subject to boundary conditions of a particular type, will have non-trivial solutions only for particular values of λ. Such solutions are the eigenfunctions of (62), and the corresponding values of λ are the eigenvalues.

Suppose now that $\varphi_m(x)$ and $\varphi_n(x)$ are two different eigenfunctions of (62) corresponding to the different eigenvalues λ_m and λ_n respectively. Then

$$\frac{d}{dx}\left(p\,\frac{d\varphi_m}{dx}\right) + (q + \lambda_m r)\varphi_m = 0 \tag{63}$$

$$\frac{d}{dx}\left(p\,\frac{d\varphi_n}{dx}\right) + (q + \lambda_n r)\varphi_n = 0. \tag{64}$$

Multiplying (63) by φ_n and (64) by φ_m and subtracting the resulting equations we have

$$(\lambda_m - \lambda_n)r\varphi_m\varphi_n = \varphi_m\,\frac{d}{dx}\left(p\,\frac{d\varphi_n}{dx}\right) - \varphi_n\,\frac{d}{dx}\left(p\,\frac{d\varphi_m}{dx}\right) \tag{65}$$

$$= \frac{d}{dx}(p\varphi_m\varphi_n' - p\varphi_n\varphi_m'), \tag{66}$$

where the primes refer to derivatives.

Integrating with respect to x between arbitrary limits a and b we have

$$(\lambda_m - \lambda_n)\int_a^b r\varphi_m\varphi_n\,dx = [p\varphi_m\varphi_n' - p\varphi_n\varphi_m']_a^b. \tag{67}$$

Consequently if

$$[p\varphi_m\varphi_n' - p\varphi_n\varphi_m']_a^b = (p\varphi_m\varphi_n' - p\varphi_n\varphi_m')_{x=b} - (p\varphi_m\varphi_n' - p\varphi_n\varphi_m')_{x=a} = 0 \tag{68}$$

then

$$(\lambda_m - \lambda_n)\int_a^b r\varphi_m\varphi_n\,dx = 0, \tag{69}$$

which shows that, since (by assumption) $\lambda_m \neq \lambda_n$, φ_m and φ_n are orthogonal on the interval (a, b) with respect to the weight function $r(x)$. Equation (68) can be satisfied in a number of ways by imposing certain types of boundary conditions on $\varphi(x)$. For example, (68) is satisfied if any of the following three conditions hold:

$$\text{(i)} \quad \varphi_m(a) = \varphi_m(b) = 0, \quad \text{for both } m \text{ and } n. \tag{70}$$

$$\text{(ii)} \quad \varphi'_m(a) = \varphi'_m(b) = 0, \quad \text{for both } m \text{ and } n. \tag{71}$$

$$\text{(iii)} \quad \varphi_m(a)\varphi'_n(a) - \varphi_n(a)\varphi'_m(a) = \varphi_m(b)\varphi'_n(b) - \varphi_n(b)\varphi'_m(b) = 0. \tag{72}$$

Furthermore if the boundary conditions are such that

$$k_1\varphi_m(a) + k_2\varphi'_m(a) = k_1\varphi_n(a) + k_2\varphi'_n(a) = 0 \tag{73}$$

and

$$l_1\varphi_m(b) + l_2\varphi'_m(b) = l_1\varphi_n(b) + l_2\varphi'_n(b) = 0 \tag{74}$$

where k_1, k_2, l_1 and l_2 are real constants, then from (73) k_1 and k_2 may take non-zero values only if the Wronskian determinant

$$\begin{vmatrix} \varphi_m(a) & \varphi_n(a) \\ \varphi'_m(a) & \varphi'_n(a) \end{vmatrix} = 0. \tag{75}$$

Likewise from (74) l_1 and l_2 may take non-zero values only if

$$\begin{vmatrix} \varphi_m(b) & \varphi_n(b) \\ \varphi'_m(b) & \varphi'_n(b) \end{vmatrix} = 0. \tag{76}$$

These two expressions, (75) and (76), are respectively

$$\varphi_m(a)\varphi'_n(a) - \varphi_n(a)\varphi'_m(a) = 0,$$
$$\varphi_m(b)\varphi'_n(b) - \varphi_n(b)\varphi'_m(b) = 0, \tag{77}$$

which together show that (72) is satisfied. Consequently boundary conditions which are linear combinations of the eigenfunctions and their derivatives at the end points of the interval (a, b) (see (73) and (74)) also give rise to eigenfunctions of the Sturm-Liouville equation which are orthogonal with respect to the weight function $r(x)$ on the interval (a, b).

Finally we note that all such boundary conditions are homogeneous in the sense that if φ is replaced by $c\varphi$, where c is an arbitrary constant, then the boundary conditions are unaltered. This would not be the case with non-homogeneous boundary conditions; for example: $\varphi_m(a) = \alpha$ (say) where α is a non-zero constant. In such cases the eigenfunctions of the Sturm-Liouville equation will not, in general, form an orthogonal set. Examples of the solution of linear partial differential equations subject to non-homogeneous boundary conditions will be given in Chapter 4, 4.3.

PROBLEMS 3

1. Show that the following sets of functions form orthogonal sets on the given intervals:

   (a) $1, \cos x, \cos 2x, \cos 3x, \ldots$ for $0 \leqslant x \leqslant \pi$.
   (b) $\sin \pi x, \sin 2\pi x, \sin 3\pi x, \ldots$ for $-1 \leqslant x \leqslant 1$.
   (c) $1, 1-x, 1-2x+\frac{1}{2}x^2$ for $0 \leqslant x < \infty$ with respect to the weight function $\omega(x) = e^{-x}$.

2. In Problem 1 (a) and (b), determine the corresponding orthonormal sets of functions, and discuss whether or not they form complete sets.

3. Show that the functions $f(x) = 1$ and $g(x) = x$ are orthogonal on the interval $-1 < x < 1$. Determine the constants α and β such that the function $h(x) = 1 + \alpha x + \beta x^2$ is orthogonal to both $f(x)$ and $g(x)$.

4. Verify the orthogonality of the first four Legendre polynomials (see 3.2) on the interval $(-1, 1)$.

5. The Hermite polynomials $H_n(x)$ are defined by

$$H_n(x) = (-1)^n e^{x^2/2} \frac{d^n}{dx^n} e^{-x^2/2}, \quad (n = 0, 1, 2, \ldots).$$

Show that

$$H_0(x) = 1, \quad H_1(x) = x, \quad H_2(x) = x^2 - 1, \quad H_3(x) = x^3 - 3x,$$

and verify by direct integration that these four functions are orthogonal on the interval $(-\infty, \infty)$ with respect to the weight function $\omega(x) = e^{-x^2/2}$.

41

CHAPTER 4

Applications of Fourier's Method

4.1 Coordinate systems and separability

In this chapter we give a few examples of the solution of linear partial differential equations using the method of separation of variables and superposition of solutions (Fourier's method: see Chapter 3, 3.1). As we have seen in 3.4 this method is applicable if, after separating the variables, the resulting ordinary differential equations are of the Sturm-Liouville type and the boundary conditions are such that the eigenfunctions form an orthogonal set. However, the question of the separability of a partial differential equation into two or more ordinary differential equations is by no means a trivial one, and clearly Fourier's method is a non-starter if the partial differential equation is found to be inseparable. As we shall see separability is closely linked with the choice of the coordinate system, the coordinates being the independent variables of the partial differential equation.

Consider, first, rectangular Cartesian coordinates (x, y, z). These coordinates are particularly suitable for the description of regions (or domains) of space which possess rectangular symmetry since the boundaries are readily defined by specifying constant values $x = a$, $y = b$, and so on. Now in this coordinate system Laplace's equation $\nabla^2 u = 0$ is separable since writing $u(x, y, z) = X(x) Y(y) Z(z)$, where X, Y and Z are functions of one variable only, we find

$$\frac{1}{XYZ} \nabla^2 u = \frac{1}{X} \frac{d^2 X}{dx^2} + \frac{1}{Y} \frac{d^2 Y}{dy^2} + \frac{1}{Z} \frac{d^2 Z}{dz^2} = 0. \tag{1}$$

Each term in the central expression of (1) is a function of one variable only, and hence each must be equal to a constant. Laplace's equation accordingly separates into the three ordinary differential equations

$$\frac{1}{X} \frac{d^2 X}{dx^2} = \alpha, \quad \frac{1}{Y} \frac{d^2 Y}{dy^2} = \beta, \quad \frac{1}{Z} \frac{d^2 Z}{dz^2} = \gamma, \tag{2}$$

where $\alpha + \beta + \gamma = 0$, α, β and γ being arbitrary constants. It is known

42

that Laplace's equation is separable in eleven different coordinate systems (see [13]) of which rectangular Cartesian coordinates (x, y, z), spherical polar coordinates (r, φ, θ), and cylindrical polar coordinates (r, φ, z) are but three. These last two systems are particularly important for the discussion of problems involving spherical and cylindrical symmetry. In order to elaborate further on the separability of partial differential equations in different coordinate systems, and to prepare the way for the solution of equations in spherically and cylindrically symmetric regions (see Chapter 5), we give here a short account of these two important coordinate systems and the corresponding forms of the Laplacian $\nabla^2 u$.

(a) *Cylindrical polar coordinates*
Let P be a typical point in space whose projection on the xy-plane is the point Q. Then if $r = OQ$, $z = QP$, and φ is the angle made by OQ with the positive x-axis, the point P is uniquely specified by the three coordinates (r, φ, z). These are the cylindrical polar coordinates of P, and are related to the rectangular Cartesian coordinates (x, y, z) of P by the equations

$$x = r \cos \varphi, \quad y = r \sin \varphi, \quad z = z \quad \text{(see Fig. 4.1)}. \tag{3}$$

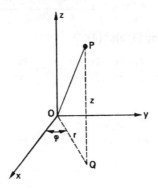

Fig. 4.1

From these equations it follows directly that

$$r = \sqrt{(x^2 + y^2)}, \quad \varphi = \tan^{-1} \frac{y}{x}. \tag{4}$$

Hence if u is a function of x, y and z (and therefore a function of

43

r, φ, z also) it follows from the elementary rules of partial differentiation that

$$\frac{\partial u}{\partial x} = \frac{\partial u}{\partial r}\frac{\partial r}{\partial x} + \frac{\partial u}{\partial \varphi}\frac{\partial \varphi}{\partial x} + \frac{\partial u}{\partial z}\frac{\partial z}{\partial x} \tag{5}$$

$$= \frac{x}{\sqrt{(x^2+y^2)}}\frac{\partial u}{\partial r} - \frac{y}{x^2+y^2}\frac{\partial u}{\partial \varphi}, \quad \text{(using (4))}, \tag{6}$$

$$= \cos \varphi \frac{\partial u}{\partial r} - \frac{\sin \varphi}{r}\frac{\partial u}{\partial \varphi}, \quad \text{(using (3))}. \tag{7}$$

Consequently the operator

$$\frac{\partial}{\partial x} = \cos \varphi \frac{\partial}{\partial r} - \frac{\sin \varphi}{r}\frac{\partial}{\partial \varphi} \tag{8}$$

and hence

$$\frac{\partial^2 u}{\partial x^2} = \frac{\partial}{\partial x}\left(\frac{\partial u}{\partial x}\right) = \left(\cos \varphi \frac{\partial}{\partial r} - \frac{\sin \varphi}{r}\frac{\partial}{\partial \varphi}\right)\left(\cos \varphi \frac{\partial u}{\partial r} - \frac{\sin \varphi}{r}\frac{\partial u}{\partial \varphi}\right) \tag{9}$$

$$= \cos^2 \varphi \frac{\partial^2 u}{\partial r^2} - \frac{2\sin \varphi \cos \varphi}{r}\frac{\partial^2 u}{\partial r\partial \varphi} + \frac{\sin^2\varphi}{r^2}\frac{\partial^2 u}{\partial \varphi^2} +$$

$$+ \frac{\sin 2\varphi}{r}\frac{\partial u}{\partial r} + \frac{2\sin \varphi \cos \varphi}{r^2}\frac{\partial u}{\partial \varphi}. \tag{10}$$

Similarly we find, using (3) and (4),

$$\frac{\partial u}{\partial y} = \frac{\partial u}{\partial r}\frac{\partial r}{\partial y} + \frac{\partial u}{\partial \varphi}\frac{\partial \varphi}{\partial y} + \frac{\partial u}{\partial z}\frac{\partial z}{\partial y} = \sin \varphi \frac{\partial u}{\partial r} + \frac{\cos \varphi}{r}\frac{\partial u}{\partial \varphi}, \tag{11}$$

and hence

$$\frac{\partial^2 u}{\partial y^2} = \frac{\partial}{\partial y}\left(\frac{\partial u}{\partial y}\right) = \left(\sin \varphi \frac{\partial}{\partial r} + \frac{\cos \varphi}{r}\frac{\partial}{\partial \varphi}\right)\left(\sin \varphi \frac{\partial u}{\partial r} + \frac{\cos \varphi}{r}\frac{\partial u}{\partial \varphi}\right) \tag{12}$$

$$= \sin^2 \varphi \frac{\partial^2 u}{\partial r^2} + \frac{2\sin \varphi \cos \varphi}{r}\frac{\partial^2 u}{\partial r\partial \varphi} + \frac{\cos^2 \varphi}{r^2}\frac{\partial^2 u}{\partial \varphi^2} +$$

$$+ \frac{\cos^2 \varphi}{r}\frac{\partial u}{\partial r} - \frac{2\sin \varphi \cos \varphi}{r^2}\frac{\partial u}{\partial \varphi}. \tag{13}$$

Hence, using (10) and (13), we find

$$\nabla^2 u \equiv \frac{\partial^2 u}{\partial x^2} + \frac{\partial^2 u}{\partial y^2} + \frac{\partial^2 u}{\partial z^2} = \frac{\partial^2 u}{\partial r^2} + \frac{1}{r}\frac{\partial u}{\partial r} + \frac{1}{r^2}\frac{\partial^2 u}{\partial \varphi^2} + \frac{\partial^2 u}{\partial z^2}, \tag{14}$$

which is the Laplacian of u in cylindrical polar coordinates. It is easily verified that an equivalent form of (14) is

$$\nabla^2 u = \frac{1}{r}\frac{\partial}{\partial r}\left(r\frac{\partial u}{\partial r}\right) + \frac{1}{r^2}\frac{\partial^2 u}{\partial \varphi^2} + \frac{\partial^2 u}{\partial z^2}. \tag{15}$$

(b) *Spherical polar coordinates*

Again let P be a typical point in space whose projection on the xy-plane is the point Q. Then if $r = OP$, φ the angle made by OQ with the positive x-axis, and θ the angle made by OP with the positive z-axis, the point P is uniquely specified by the three coordinates (r, φ, θ). These are the spherical polar coordinates of P, and are related to its rectangular Cartesian coordinates by the relations

$$x = r\sin\theta\cos\varphi, \quad y = r\sin\theta\sin\varphi, \quad z = r\cos\theta \quad \text{(see Fig. 4.2)}. \tag{16}$$

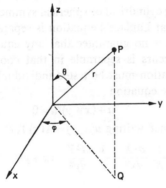

Fig. 4.2

Following a similar analysis to that of (a) it can be shown that in spherical polar coordinates

$$\nabla^2 u = \frac{\partial^2 u}{\partial r^2} + \frac{2}{r}\frac{\partial u}{\partial r} + \frac{1}{r^2\sin^2\theta}\frac{\partial^2 u}{\partial \varphi^2} + \frac{1}{r^2}\frac{\partial^2 u}{\partial \theta^2} + \frac{\cot\theta}{r^2}\frac{\partial u}{\partial \theta} \tag{17}$$

$$= \frac{1}{r^2}\left[\frac{\partial}{\partial r}\left(r^2\frac{\partial u}{\partial r}\right) + \frac{1}{\sin^2\theta}\frac{\partial^2 u}{\partial \varphi^2} + \frac{1}{\sin\theta}\frac{\partial}{\partial \theta}\left(\sin\theta\frac{\partial u}{\partial \theta}\right)\right]. \tag{18}$$

With the results of (a) and (b) we can now demonstrate, for example, that Laplace's equation $\nabla^2 u = 0$ is separable in cylindrical

45

polar coordinates. For writing in (15) $u(r, \varphi, z) = R(r)\Phi(\varphi)Z(z)$ we find

$$\nabla^2 u = \frac{1}{rR} \frac{d}{dr}\left(r \frac{dR}{dr}\right) + \frac{1}{r^2\Phi} \frac{d^2\Phi}{d\varphi^2} + \frac{1}{Z} \frac{d^2Z}{dz^2} = 0, \qquad (19)$$

whence

$$\frac{1}{Z} \frac{d^2Z}{dz^2} = \alpha, \quad \frac{1}{\Phi} \frac{d^2\Phi}{d\varphi^2} = -\beta, \qquad (20)$$

where α and β are arbitrary constants. We therefore obtain

$$\frac{1}{r} \frac{d}{dr}\left(r \frac{dR}{dr}\right) + \left(\alpha - \frac{\beta}{r^2}\right)R = 0. \qquad (21)$$

Accordingly Laplace's equation separates into the three ordinary differential equations (20) and (21). Equation (21) is known as Bessel's equation and its solutions – the Bessel functions – arise frequently in the solutions of partial differential equations in regions of space possessing cylindrical or spherical symmetry (see Chapter 5).

Now the fact that Laplace's equation is separable in a particular coordinate system is no guarantee that any equation in which the Laplacian $\nabla^2 u$ occurs is separable in that coordinate system. In general, every equation must be tested individually for separability.

For example, the equation

$$\nabla^2 u + (x + y)^2 u = 0 \qquad (22)$$

is not separable since writing $u(x, y) = X(x)Y(y)$ we find

$$\frac{1}{X} \frac{d^2X}{dx^2} + \frac{1}{Y} \frac{d^2Y}{dy^2} + (x + y)^2 = 0 \qquad (23)$$

or

$$\left(\frac{1}{X} \frac{d^2X}{dx^2} + x^2\right) + \left(\frac{1}{Y} \frac{d^2Y}{dy^2} + y^2\right) + 2xy = 0. \qquad (24)$$

Although the terms in brackets are each functions of one variable only, this is not true of the last term.

The equation

$$\nabla^2 u + (x^2 + y^2)^2 u = 0 \qquad (25)$$

likewise is not separable as it stands, but when written in terms of polar coordinates (r, φ) (the two-dimensional form of (a)) becomes using (15) and (3)

$$\frac{1}{r} \frac{\partial}{\partial r}\left(r \frac{\partial u}{\partial r}\right) + \frac{1}{r^2} \frac{\partial^2 u}{\partial \varphi^2} + r^4 u = 0. \qquad (26)$$

Writing $u(r, \varphi) = R(r)\Phi(\varphi)$ we find

$$\frac{1}{R}\left[r\frac{d}{dr}\left(r\frac{dR}{dr}\right) + r^6 R\right] + \frac{1}{\Phi}\frac{d^2\Phi}{d\varphi^2} = 0. \tag{27}$$

Hence in these coordinates (25) separates into the pair of ordinary differential equations

$$r\frac{d}{dr}\left(r\frac{dR}{dr}\right) + (r^6 + \lambda)R = 0, \quad \frac{d^2\Phi}{d\varphi^2} = \lambda\Phi, \tag{28}$$

where λ is an arbitrary constant.

4.2 Homogeneous equations

The following examples all refer to the simpler equations of mathematical physics as discussed in Chapter 1. Although we shall not be concerned with the explicit physical interpretation in each case, nevertheless the coordinates (x, y, z) are thought of as space coordinates and t as a time variable.

Example 1. To obtain the solution of the equation

$$\frac{\partial^2 u}{\partial x^2} = \frac{1}{k}\frac{\partial u}{\partial t}, \tag{29}$$

where k is a constant, satisfying the boundary conditions

$$u(0, t) = u(l, t) = 0, \quad t \geqslant 0, \tag{30}$$
$$u(x, 0) = f(x), \quad 0 \leqslant x \leqslant l, \tag{31}$$

f being a given function and l a constant.
Assuming a solution of the form

$$u(x, t) = X(x)T(t) \tag{32}$$

and substituting in (29) we find

$$\frac{1}{X}\frac{d^2 X}{dx^2} = \frac{1}{kT}\frac{dT}{dt}. \tag{33}$$

The expression on the left depends only on x, whilst the expression on the right is a function of t only. As in 3.1 we deduce that each must be equal to a constant λ, say. For $\lambda \geqslant 0$ we may show that the only solution of the type (32) consistent with the boundary conditions (30) is $u(x, t) \equiv 0$. For negative λ $(= -\omega^2$, say) we have

$$\frac{1}{X}\frac{d^2 X}{dx^2} = -\omega^2, \tag{34}$$

$$\frac{1}{kT}\frac{dT}{dt} = -\omega^2, \tag{35}$$

whence
$$X(x) = A \cos \omega x + B \sin \omega x, \tag{36}$$
and
$$T(t) = Ce^{-\omega^2 kt}, \tag{37}$$
where A, B and C are arbitrary constants.

With (36) and (37), (32) becomes
$$u(x, t) = (A \cos \omega x + B \sin \omega x)e^{-\omega^2 kt}, \tag{38}$$
the constant C having been absorbed into A and B without loss of generality. Putting $x = 0$ in (38) and using (30) we have
$$0 = Ae^{-\omega^2 kt}, \tag{39}$$
for all t, which implies
$$A = 0. \tag{40}$$
Secondly, putting $x = l$ in (38), we find using (30) and (40) that
$$0 = (B \sin \omega l)e^{-\omega^2 kt}, \tag{41}$$
which, since we cannot put $B = 0$ without making $u(x, t) \equiv 0$, leads to non-trivial (i.e. not identically zero) solutions provided
$$\sin \omega l = 0. \tag{42}$$
Hence
$$\omega_r = \frac{r\pi}{l} \quad \text{where } r = 1, 2, 3, \ldots, \tag{43}$$
(the case $r = 0$ being excluded to avoid making $u(x, t) \equiv 0$ again). From (40), (43) and (38) we now obtain an infinity of eigenvalues and eigenfunctions

$$\omega_1 = \frac{\pi}{l}, \quad u_1(x, t) = B_1 e^{-\pi^2 kt/l^2} \sin \frac{\pi x}{l},$$

$$\omega_2 = \frac{2\pi}{l}, \quad u_2(x, t) = B_2 e^{-4\pi^2 kt/l^2} \sin \frac{2\pi x}{l}, \tag{44}$$

$$\cdot \qquad \cdot \qquad \cdot \qquad \cdot$$
$$\cdot \qquad \cdot \qquad \cdot \qquad \cdot$$
$$\cdot \qquad \cdot \qquad \cdot \qquad \cdot$$

$$\omega_r = \frac{r\pi}{l}, \quad u_r(x, t) = B_r e^{-r^2\pi^2 kt/l^2} \sin \frac{r\pi x}{l},$$

$$\cdot \qquad \cdot \qquad \cdot \qquad \cdot$$
$$\cdot \qquad \cdot \qquad \cdot \qquad \cdot$$
$$\cdot \qquad \cdot \qquad \cdot \qquad \cdot$$

and so on, B_1, B_2, ..., B_r, ... being arbitrary constants. We now take the linear combination of the solutions (44)

$$u(x, t) = \sum_{r=1}^{\infty} B_r e^{-r^2\pi^2 kt/l^2} \sin \frac{r\pi x}{l} \tag{45}$$

as the general solution of (29) satisfying the boundary conditions (30), where the constants B_r must be chosen to satisfy the remaining boundary condition (31). Hence putting $t = 0$ in (45) we have

$$f(x) = \sum_{r=1}^{\infty} B_r \sin \frac{r\pi x}{l}, \tag{46}$$

from which it follows, using the orthonormality properties of the sine function (see Chapter 3, equation (38)), that

$$B_r = \frac{2}{l} \int_0^l f(x) \sin \frac{r\pi x}{l} \, dx, \quad (r = 1, 2, 3, \dots). \tag{47}$$

The solution of (29) subject to the given boundary conditions is therefore

$$u(x, t) = \sum_{r=1}^{\infty} \left\{ \left[\frac{2}{l} \int_0^l f(x') \sin \frac{r\pi x'}{l} \, dx' \right] e^{-r^2\pi^2 kt/l^2} \sin \frac{r\pi x}{l} \right\}, \tag{48}$$

where x' has been written as the variable of integration to avoid confusion with the independent variable x.

As a specific example we consider the case where

$$f(x) = \begin{cases} x & \text{for } 0 \leqslant x \leqslant \dfrac{l}{2}, \\ l - x & \text{for } \dfrac{l}{2} \leqslant x \leqslant l. \end{cases} \tag{49}$$

Using (47) the reader may easily verify that

$$B_r = \begin{cases} \dfrac{4l}{r^2\pi^2}, & r = 1, 5, 9, \dots \\ \dfrac{-4l}{r^2\pi^2}, & r = 3, 7, 11, \dots \\ 0, & r \text{ even.} \end{cases} \tag{50}$$

Hence finally

$$u(x, t) = \frac{4l}{\pi^2} \left(e^{-\pi^2 kt/l^2} \sin \frac{\pi x}{l} - \frac{1}{9} e^{-9\pi^2 kt/l^2} \sin \frac{3\pi x}{l} + \dots \right). \tag{51}$$

Example 2. If instead of (30) we impose different boundary conditions on (29) then different eigenvalues and eigenfunctions will arise. For example, suppose instead of (30) we now require

$$\left(\frac{\partial u}{\partial x}\right)_{x=0} = \left(\frac{\partial u}{\partial x}\right)_{x=l} = 0, \quad t \geqslant 0, \qquad (52)$$

leaving (31) unchanged. Then proceeding as before the eigenfunctions are

$$\omega_r = \frac{r\pi}{l}, \quad u_r(x, t) = A_r e^{-r^2\pi^2 kt/l^2} \cos \frac{r\pi x}{l}, \qquad (53)$$

where $r = 0, 1, 2, 3, \ldots$

Hence

$$u(x, t) = A_0 + \sum_{r=1}^{\infty} A_r e^{-r^2\pi^2 kt/l^2} \cos \frac{r\pi x}{l}. \qquad (54)$$

At $t = 0$, $u(x, 0) = f(x)$. Consequently

$$f(x) = A_0 + \sum_{r=1}^{\infty} A_r \cos \frac{r\pi x}{l}, \qquad (55)$$

and accordingly using Fourier techniques (see Chapter 3, 3.3)

$$A_0 = \frac{1}{l} \int_0^l f(x)dx, \quad A_r = \frac{2}{l} \int_0^l f(x) \cos \frac{r\pi x}{l} \, dx. \qquad (56)$$

These values when inserted in (54) give the required solution. We note that as $t \to \infty$

$$u(x, t) \to \frac{1}{l} \int_0^l f(x)dx, \qquad (57)$$

which is just the mean value of $f(x)$ on the interval $(0, l)$.

Example 3. To obtain the solution of the two-dimensional Laplace equation

$$\nabla^2 u = \frac{\partial^2 u}{\partial x^2} + \frac{\partial^2 u}{\partial y^2} = 0 \qquad (58)$$

within the rectangle R defined by $0 \leqslant x \leqslant a$, $0 \leqslant y \leqslant b$, where a and b are constants, which satisfies the Dirichlet conditions

$$u(x, y) = \begin{cases} 0 & \text{when } x = 0, \quad 0 \leqslant y \leqslant b, \\ 0 & \text{when } x = a, \quad 0 \leqslant y \leqslant b, \\ 0 & \text{when } y = b, \quad 0 \leqslant x \leqslant a, \\ f(x) & \text{when } y = 0, \quad 0 < x < a. \end{cases} \qquad (59)$$

Writing

$$u(x, y) = X(x)Y(y) \tag{60}$$

(58) becomes

$$\frac{1}{X}\frac{d^2 X}{dx^2} + \frac{1}{Y}\frac{d^2 Y}{dy^2} = 0 \tag{61}$$

whence

$$\frac{1}{X}\frac{d^2 X}{dx^2} = \lambda, \quad \frac{1}{Y}\frac{d^2 Y}{dy^2} = -\lambda, \tag{62}$$

where λ is an arbitrary constant. When $\lambda = 0$, the solutions of (62) are respectively

$$X = Ax + B, \quad Y = Cy + D \tag{63}$$

giving

$$u(x, y) = (Ax + B)(Cy + D), \tag{64}$$

where A, B, C and D are arbitrary constants. Imposing the first two boundary conditions of (59) leads to $u(x, y) \equiv 0$. Likewise when $\lambda > 0$ ($= \omega^2$, say) the solution has the form

$$u(x, y) = (A \cosh \omega x + B \sinh \omega x)(C \cos \omega y + D \sin \omega y) \tag{65}$$

which with the first pair of boundary conditions again leads to $u(x, y) \equiv 0$. However, when $\lambda < 0$ ($= -\omega^2$) the solutions are

$$X = A \cos \omega x + B \sin \omega x, \quad Y = C \cosh \omega y + D \sinh \omega y \tag{66}$$

giving

$$u(x, y) = (A \cos \omega x + B \sin \omega x)(C \cosh \omega y + D \sinh \omega y). \tag{67}$$

Imposing the first boundary condition of (59) we have

$$0 = A(C \cosh \omega y + D \sinh \omega y) \quad (0 \leqslant y \leqslant b) \tag{68}$$

whence

$$A = 0. \tag{69}$$

Similarly the second boundary condition leads to

$$B \sin \omega a = 0, \tag{70}$$

which yields for non-trivial solutions

$$\omega = \frac{r\pi}{a}, \quad \text{where } r = 1, 2, 3, \ldots \tag{71}$$

Likewise the third boundary condition of (59) gives (putting $y = b$ in (67))

$$\frac{C}{D} = -\tanh \omega b. \tag{72}$$

Hence substituting (69), (71) and (72) into (67) we find the eigenvalue-eigenfunction system

$$\omega_1 = \frac{\pi}{a}, \quad u_1(x, y) = E_1 \sin \frac{\pi x}{a} \sinh \frac{\pi(b-y)}{a},$$

$$\omega_2 = \frac{2\pi}{a}, \quad u_2(x, y) = E_2 \sin \frac{2\pi x}{a} \sinh \frac{2\pi(b-y)}{a}, \qquad (73)$$

$$\cdot \qquad \cdot \qquad \cdot \qquad \cdot$$
$$\cdot \qquad \cdot \qquad \cdot \qquad \cdot$$
$$\cdot \qquad \cdot \qquad \cdot \qquad \cdot$$

$$\omega_r = \frac{r\pi}{a}, \quad u_r(x, y) = E_r \sin \frac{r\pi x}{a} \sinh \frac{r\pi(b-y)}{a},$$

$$\cdot \qquad \cdot \qquad \cdot \qquad \cdot$$
$$\cdot \qquad \cdot \qquad \cdot \qquad \cdot$$
$$\cdot \qquad \cdot \qquad \cdot \qquad \cdot$$

where $E_1, E_2, \ldots, E_r, \ldots$, are arbitrary constants.

As before we now take a linear combination of these solutions to get

$$u(x, y) = \sum_{r=1}^{\infty} E_r \sin \frac{r\pi x}{a} \sinh \frac{r\pi(b-y)}{a} \qquad (74)$$

as the general solution of (58) satisfying the first boundary conditions of (59). The constants E_r in this solution must now be chosen to satisfy the fourth and last boundary condition of (59). Hence putting $y = 0$ in (74)

$$f(x) = \sum_{r=1}^{\infty} E_r \sinh \frac{r\pi b}{a} \sin \frac{r\pi x}{a} \qquad (75)$$

from which we find, using the orthonormality of the sine function on the interval $(0, a)$,

$$E_r \sinh \frac{r\pi b}{a} = \frac{2}{a} \int_0^a f(x) \sin \frac{r\pi x}{a} \, dx. \qquad (76)$$

Consequently substituting (76) into (74) the final solution of (58) satisfying (59) is

$$u(x, y) = \sum_{r=1}^{\infty} \left\{ \left[\frac{2}{a} \int_0^a f(x') \sin \frac{r\pi x'}{a} \, dx' \right] \frac{\sin \dfrac{r\pi x}{a} \sinh \dfrac{r\pi(b-y)}{a}}{\sinh \dfrac{r\pi b}{a}} \right\}. \qquad (77)$$

Example 4. To obtain the eigenvalues of the biharmonic wave equation

$$\frac{\partial^4 u}{\partial x^4} = -\frac{1}{c^2}\frac{\partial^2 u}{\partial t^2}, \tag{78}$$

where c is a constant, subject to the boundary conditions

$$u(0, t) = u(l, t) = 0, \quad t \geqslant 0, \tag{79}$$

$$\left(\frac{\partial u}{\partial x}\right)_{x=0} = \left(\frac{\partial u}{\partial x}\right)_{x=l} = 0, \quad t \geqslant 0, \tag{80}$$

where l is a constant.

Writing $u(x, t) = X(x)T(t)$ and proceeding as in the earlier example, we find that non-trivial solutions ($u(x, t) \not\equiv 0$) satisfying (79) and (80) exist only if

$$\frac{1}{X}\frac{d^4 X}{dx^4} = \lambda^4, \quad \frac{1}{T}\frac{d^2 T}{dt^2} = -\lambda^4 c^2, \tag{81}$$

where λ is an arbitrary real non-zero constant. From these two equations we find

$$X(x) = A\cosh \lambda x + B\cos \lambda x + C\sinh \lambda x + D\sin \lambda x, \tag{82}$$

$$T(t) = E\cos \lambda^2 ct + F\sin \lambda^2 ct. \tag{83}$$

Applying the pair of boundary conditions at $x = 0$ we have $X(0) = X'(0) = 0$, and from (82) and its derivative therefore

$$B = -A, \quad D = -C. \tag{84}$$

Hence

$$X(x) = A(\cosh \lambda x - \cos \lambda x) + C(\sinh \lambda x - \sin \lambda x). \tag{85}$$

At $x = l$ we have $X(l) = X'(l) = 0$ and consequently from (85) and its derivative

$$0 = A(\cosh \lambda l - \cos \lambda l) + C(\sinh \lambda l - \sin \lambda l), \tag{86}$$

$$0 = A\lambda(\sinh \lambda l + \sin \lambda l) + C\lambda(\cosh \lambda l - \cos \lambda l). \tag{87}$$

Since we cannot have $\lambda = 0$ without producing the trivial solution $u(x, t) \equiv 0$, we may cancel λ from the last equation. Equations (86) and (87) are then consistent only if the ratio A/C is the same for each. This is equivalent to requiring the determinant

$$\begin{vmatrix} \cosh \lambda l - \cos \lambda l & \sinh \lambda l - \sin \lambda l \\ \sinh \lambda l + \sin \lambda l & \cosh \lambda l - \cos \lambda l \end{vmatrix} = 0. \tag{88}$$

53

Straightforward expansion yields

$$\cosh \lambda l \cos \lambda l = 1, \tag{89}$$

which is the equation for the eigenvalues λ. The roots of this equation can be calculated numerically, the first five being

$$\lambda_1 l = 0, \quad \lambda_2 l = 4 \cdot 73, \quad \lambda_3 l = 7 \cdot 85, \quad \lambda_4 l = 11 \cdot 00, \quad \lambda_5 l = 14 \cdot 14, \ldots \tag{90}$$

The first eigenvalue which is zero is of no importance since, as we have already remarked, $\lambda = 0$ gives $u(x, t) \equiv 0$.

4.3 Non-homogeneous boundary conditions

In all the examples of 4.2 the boundary conditions at fixed values of x have been homogeneous (i.e. taking zero values). Such boundary conditions as we have seen in Chapter 3, 3.4 are of the Sturm-Liouville type and enable the solution of the boundary value problem to be obtained immediately in terms of orthogonal sets of functions.

We now discuss by means of two examples the method to be adopted when non-homogeneous boundary conditions are encountered on the space variable x. For simplicity we consider problems involving only one space dimension.

Example 5. To obtain a solution of the equation

$$\frac{\partial^2 u}{\partial x^2} = \frac{1}{k} \frac{\partial u}{\partial t} \tag{91}$$

subject to the boundary conditions

$$u(0, t) = U_0, \quad t \geqslant 0, \tag{92}$$

$$u(l, t) = U_1, \quad t \geqslant 0, \tag{93}$$

and

$$u(x, 0) = f(x), \quad 0 \leqslant x \leqslant l, \tag{94}$$

where U_0, U_1, k and l are given constants and $f(x)$ is a given function. This problem is similar to Example 1 of 4.2 except for the appearance of constants on the right-hand sides of (92) and (93) instead of zeros (see (30)).

We write now

$$u(x, t) = v(x) + \omega(x, t), \tag{95}$$

where $v(x)$ is a time-independent function representing the 'stationary state' solution, and $\omega(x, t)$ the function representing the deviation from that stationary state. Inserting (95) into (91) we find that the functions $v(x)$ and $\omega(x, t)$ satisfy respectively

$$\frac{d^2v}{dx^2} = 0, \tag{96}$$

and

$$\frac{1}{k} \frac{\partial \omega}{\partial t} = \frac{\partial^2 \omega}{\partial x^2}, \tag{97}$$

together with the boundary conditions (using (92), (93), (94) and 95))

$$v(0) = U_0, \quad v(l) = U_1, \tag{98}$$

and

$$\omega(0, t) = 0, \quad \omega(l, t) = 0, \quad t \geqslant 0, \tag{99}$$

$$\omega(x, 0) = f(x) - v(x), \quad 0 \leqslant x \leqslant l. \tag{100}$$

Hence solving for $v(x)$ using (96) and (98) we find

$$v(x) = U_0 + \frac{x}{l}(U_1 - U_0). \tag{101}$$

The function $\omega(x, t)$ is now to be obtained from (97) and the boundary conditions (99) now being of the homogeneous type. Accordingly $\omega(x, t)$ is just the solution given in Example 1 of 4.2, except for the alteration in the function on the right-hand side of (100). The solution of (91) subject to (92)–(94) is therefore

$$u(x, t) = U_0 + \frac{x}{l}(U_1 - U_0) + \sum_{r=1}^{\infty} \bar{B}_r e^{-r^2\pi^2kt/l^2} \sin \frac{r\pi x}{l}, \tag{102}$$

where

$$\bar{B}_r = \frac{2}{l} \int_0^l [f(x) - v(x)] \sin \frac{r\pi x}{l} \, dx \tag{103}$$

$$= \frac{2}{l} \int_0^l \left[f(x) - U_0 - \frac{x}{l}(U_1 - U_0) \right] \sin \frac{r\pi x}{l} \, dx \tag{104}$$

$$= \frac{2}{l} \int_0^l f(x) \sin \frac{r\pi x}{l} \, dx + \frac{2}{r\pi} [(-1)^r U_1 - U_0], \tag{105}$$

where $r = 1, 2, 3, \ldots$

When U_0 and U_1 are zero (105) reduces to the solution (48) as required.

A completely similar procedure may be applied to other equations subject to non-homogeneous boundary conditions. It is left to the

55

reader to carry through the analysis for the boundary value problem

$$\frac{\partial^2 u}{\partial x^2} = \frac{1}{c^2} \frac{\partial^2 u}{\partial t^2} \tag{106}$$

subject to

$$u(0, t) = U_0, \quad u(l, t) = U_1, \quad t \geqslant 0, \tag{107}$$

$$u(x, 0) = f(x), \quad \left[\frac{\partial u}{\partial t}\right]_{t=0} = g(x), \quad 0 \leqslant x \leqslant l, \tag{108}$$

where U_0, U_1, l and c are constants and $f(x)$ and $g(x)$ are given functions (see Example 1 of Chapter 3).

Example 6. We now consider the case where U_1 and U_0 are no longer constants but are given functions of time. Suppose we wish to solve

$$\frac{\partial^2 u}{\partial x^2} = \frac{1}{k} \frac{\partial u}{\partial t} \tag{109}$$

subject to the boundary conditions

$$u(0, t) = U_0(t), \quad t \geqslant 0, \tag{110}$$
$$u(l, t) = U_1(t), \quad t \geqslant 0, \tag{111}$$

and

$$u(x, 0) = f(x), \tag{112}$$

where as usual k and l are given constants, and $U_0(t)$, $U_1(t)$ and $f(x)$ are given functions.

We now write

$$u(x, t) = v(x, t) + \omega(x, t) \tag{113}$$

which with (109) gives

$$\frac{\partial^2 \omega}{\partial x^2} - \frac{1}{k} \frac{\partial \omega}{\partial t} = -\left[\frac{\partial^2 v}{\partial x^2} - \frac{1}{k} \frac{\partial v}{\partial t}\right] \tag{114}$$

together with the boundary conditions (using (110) and (111)),

$$\omega(0, t) = U_0(t) - v(0, t), \quad t \geqslant 0, \tag{115}$$
$$\omega(l, t) = U_1(t) - v(l, t), \quad t \geqslant 0, \tag{116}$$

and

$$\omega(x, 0) = f(x) - v(x, 0), \quad 0 \leqslant x \leqslant l. \tag{117}$$

56

The function $v(x, t)$ is now to be chosen so that the boundary conditions (115) and (116) become homogeneous (i.e. right-hand sides equal to zero) since this then opens the way to the usual method of solution for $\omega(x, t)$. Clearly choosing

$$v(x, t) = U_0(t) + \frac{x}{l} [U_1(t) - U_0(t)] \tag{118}$$

meets these requirements. The equation for $\omega(x, t)$ (see (114)) now becomes

$$\frac{\partial^2 \omega}{\partial x^2} - \frac{1}{k} \frac{\partial \omega}{\partial t} = \frac{1}{k} \left[U_0'(t) + \frac{x}{l} (U_1'(t) - U_0'(t)) \right], \tag{119}$$

where the primes refer to the first derivatives with respect to t. This equation is inhomogeneous and is a special case of the generalised heat conduction

$$\frac{\partial^2 \omega}{\partial x^2} - \frac{1}{k} \frac{\partial \omega}{\partial t} = F(x, t), \tag{120}$$

where $F(x, t)$ is (usually) a known function.

The solutions of inhomogeneous equations can be obtained in a number of different ways (see, for example, the next section, Chapter 8, 8.4, and Chapter 9). However, the important point here is that when non-homogeneous boundary conditions, which are functions of t, are imposed on the *homogeneous* equation (109) they lead ultimately to the need to solve an *inhomogeneous* equation. This situation did not arise in the previous example (Example 5) where the non-homogeneous boundary values were constants. Inhomogeneous equations (such as Poisson's equation – see Chapter 1, 1.3 and Chapter 8, Example 11) are of particular importance in mathematical physics, and it is clear from this last example that they can arise in most unexpected ways.

4.4 Inhomogeneous equations
We now discuss one method of solving inhomogeneous equations which uses Fourier series expansions. Consider the equation

$$\frac{\partial u}{\partial t} = \frac{1}{k} \frac{\partial^2 u}{\partial x^2} + f(x, t) \tag{121}$$

where $u \equiv u(x, t)$, k is a constant and $f(x, t)$ a given function (compare

57

with (120)). Suppose (121) is to be solved for $0 < x < l$, $t > 0$ (where l is a given constant), subject to the boundary conditions

$$u(0, t) = 0, \quad u(l, t) = 0, \tag{122}$$

and

$$u(x, 0) = \varphi(x), \tag{123}$$

where $\varphi(x)$ is a given function.

We first write

$$u(x, t) = \sum_{r=1}^{\infty} u_r(t) \sin \frac{r\pi x}{l}, \tag{124}$$

which ensures that the boundary conditions (122) are satisfied. Suppose now we expand $f(x, t)$ in a similar way so that

$$f(x, t) = \sum_{r=1}^{\infty} f_r(t) \sin \frac{r\pi x}{l}, \tag{125}$$

then using the orthonormality property of the sine functions (see Chapter 3, equation (38)) we deduce from (125) by multiplying each side by $\sin \frac{s\pi x}{l}$ and integrating with respect to x from $0 \rightarrow l$ that

$$f_r(t) = \frac{2}{l} \int_0^l f(\xi, t) \sin \frac{r\pi \xi}{l} \, d\xi, \tag{126}$$

where ξ is a parameter of integration.

Inserting the series (124) and (125) into (121) we find

$$\sum_{r=1}^{\infty} \left\{ \left[\frac{d}{dt} u_r(t) + \frac{1}{k} \left(\frac{r\pi}{l} \right)^2 u_r(t) - f_r(t) \right] \sin \frac{r\pi x}{l} \right\} = 0, \tag{127}$$

which leads immediately to the first-order ordinary differential equation

$$\frac{d}{dt} u_r(t) + \frac{1}{k} \left(\frac{r\pi}{l} \right)^2 u_r(t) = f_r(t) \tag{128}$$

for $u_r(t)$, $f_r(t)$ being known from (126). The boundary condition to be imposed on this equation is determined by putting $t = 0$ in (124), using (123), and expanding $\varphi(x)$ as a Fourier sine series

$$\varphi(x) = \sum_{r=1}^{\infty} B_r \sin \frac{r\pi x}{l}, \tag{129}$$

where, as usual,

$$B_r = \frac{2}{l} \int_0^l \varphi(\xi) \sin \frac{r\pi \xi}{l} \, d\xi. \tag{130}$$

In this way we have

$$u(x, 0) = \sum_{r=1}^{\infty} u_r(0) \sin \frac{r\pi x}{l} = \varphi(x) = \sum_{r=1}^{\infty} B_r \sin \frac{r\pi x}{l}. \qquad (131)$$

Hence, comparing the second and fourth terms of (131), we find

$$u_r(0) = B_r. \qquad (132)$$

where B_r is a constant whose value is given by (130).

Accordingly $u_r(t)$ is determined uniquely by solving (128) subject to (132). This solution when inserted into (124) leads to the required solution $u(x, t)$ of (121).

The validity of this method (which may frequently be applied to other types of linear inhomogeneous equations) clearly depends on the assumption that the necessary functions are all expandable as Fourier series.

PROBLEMS 4

1. Obtain all solutions of the equation

$$\frac{\partial^2 u}{\partial x^2} - \frac{\partial u}{\partial y} = u$$

of the form $u(x, y) = (A \cos \lambda x + B \sin \lambda x)f(y)$, where A, B and λ are constants. Find a solution of the equation for which $u = 0$ when $x = 0$; $u = 0$ when $x = \pi$, $u = x$ when $y = 1$.

2. Show that the solution of the equation

$$x^2 \frac{\partial^2 u}{\partial x^2} = \frac{1}{c^2} \frac{\partial^2 u}{\partial t^2} \quad (c = \text{constant})$$

which satisfies the boundary condition $u = 0$ for $x = a$ and $x = 2a$, is

$$u(x, t) = \sin\left[\lambda \log_e \left(\frac{x}{a}\right)\right]\left(\frac{x}{a}\right)^{1/2}(A \cos \omega t + B \sin \omega t),$$

where $\omega^2 = c^2(\lambda^2 + \frac{1}{4})$, and $\lambda = n\pi/\log_e 2$, n being a positive integer.

3. A uniform string of length πl is fastened at its ends $x = 0$, $x = \pi l$. The point $x = \frac{1}{3}\pi l$ is drawn aside a small distance b, and the string released from rest at time $t = 0$. Show that, if T is the tension and

ρ the density per unit length of string, the subsequent displacement of the string is

$$u(x, t) = \frac{9b}{\pi^2} \sum_{n=1}^{\infty} \frac{1}{n^2} \sin \frac{n\pi}{3} \sin \frac{nx}{l} \cos \frac{nct}{l},$$

where $c^2 = T/\rho$.

4. Show that the solution of the equation

$$\frac{\partial^2 u}{\partial x^2} = \frac{1}{k} \frac{\partial u}{\partial t}$$

which satisfies the conditions:

(i) $\dfrac{\partial u}{\partial x} = 0$ where $x = 0$ and $x = a$, for all t,

(ii) u is bounded for $-a \leqslant x \leqslant a$, as $t \to \infty$,

(iii) $u = |x|$ for $-a \leqslant x \leqslant a$ when $t = 0$,

is

$$y = \tfrac{1}{2}a - \frac{4a}{\pi^2} \sum_{n=0}^{\infty} \frac{1}{(2n+1)^2} \cos \frac{(2n+1)\pi x}{a} \, e^{-[(2n+1)^2 k\pi^2 t]/a^2}.$$

5. Given the equation

$$\frac{\partial^2 u}{\partial x^2} = \frac{1}{k} \frac{\partial u}{\partial t} + \frac{\beta}{k}(u - u_0),$$

where k, β and u_0 are constants, and the boundary conditions

$$u(x, 0) = f(x), \quad u(0, t) = u_1, \quad u(l, t) = u_2,$$

where u_1 and u_2 are constants, show by means of the substitution

$$u(x, t) = u_0 + v(x, t)e^{-\beta t}$$

that the equation reduces to

$$\frac{\partial^2 v}{\partial x^2} = \frac{1}{k} \frac{\partial v}{\partial t}$$

and the boundary conditions to

$$v(x, 0) = f(x) - u_0, \quad v(0, t) = (u_1 - u_0)e^{\beta t}, \quad v(l, t) = (u_2 - u_0)^{\beta t}.$$

Hence show that when $u_1 = u_2 = u_0$

$$u(x, t) = u_0 + \sum_{n=1}^{\infty} b_n \sin \frac{n\pi x}{l} \, e^{-[(\beta l^2 + n^2\pi^2 k)t]/l^2}$$

60

where

$$b_n = \frac{2}{l} \int_0^l f(x) \sin \frac{n\pi x}{l} \, dx - \frac{2u_0}{n\pi} (1 - \cos n\pi).$$

6. Show that the solution of the equation

$$\frac{\partial^2 u}{\partial x^2} = \frac{1}{k} \frac{\partial u}{\partial t}$$

which satisfies the conditions

$$u = 0 \text{ at } x = 0, \quad \text{for } t > 0; \quad u = 1 \text{ at } x = 1, \quad \text{for } t > 0$$

and $\quad u = 0$ at $t = 0$, for $0 \leqslant x \leqslant 1$, is

$$u(x, t) = x + \frac{2}{\pi} \sum_{n=1}^{\infty} \frac{(-1)^n}{n} e^{-n^2\pi^2kt}. \sin n\pi x.$$

CHAPTER 5

Problems involving Cylindrical and Spherical Symmetry

5.1 Simple solutions of Laplace's equation

(a) *Cylindrical symmetry.*

Using the form of the Laplacian in cylindrical polar coordinates (r, φ, z) derived in Chapter 4 (equation (15)), Laplace's equation is

$$\nabla^2 u = \frac{1}{r} \frac{\partial}{\partial r}\left(r \frac{\partial u}{\partial r}\right) + \frac{1}{r^2} \frac{\partial^2 u}{\partial \varphi^2} + \frac{\partial^2 u}{\partial z^2} = 0. \tag{1}$$

Restricting ourselves to two-dimensions so that there is no z dependence, and assuming circular symmetry (i.e. no dependence on the angular coordinate φ – see Fig. 4.1), we have

$$\nabla^2 u = \frac{1}{r} \frac{d}{dr}\left(r \frac{du}{dr}\right) = 0, \tag{2}$$

where $u = u(r)$.

Hence solving (2) we find

$$u(r) = A \log_e r + B, \tag{3}$$

where A and B are arbitrary constants. We note that the function $v(r) = u\left(\frac{1}{r}\right)$ is also a harmonic function since

$$u\left(\frac{1}{r}\right) = A \log_e \frac{1}{r} + B = A \log_e 1 - A \log_e r + B = A' \log_e r + B \tag{4}$$

which has the same form as (3). The solution (3) is singular at both $r = 0$ and at infinity provided $A \neq 0$.

(b) *Spherical symmetry.*

In spherical polar coordinates (r, φ, θ) Laplace's equation takes the form (see Chapter 4, equation (18))

$$\nabla^2 u = \frac{1}{r^2}\left[\frac{\partial}{\partial r}\left(r^2 \frac{\partial u}{\partial r}\right) + \frac{1}{\sin^2 \theta} \frac{\partial^2 u}{\partial \varphi^2} + \frac{1}{\sin \theta} \frac{\partial}{\partial \theta}\left(\sin \theta \frac{\partial u}{\partial \theta}\right)\right] = 0. \tag{5}$$

62

The specialisation to spherical symmetry is obtained by taking u as a function of r only. Then (5) becomes

$$\frac{d}{dr}\left(r^2\frac{du}{dr}\right)=0,\qquad(6)$$

which has the solution

$$u(r)=\frac{A}{r}+B,\qquad(7)$$

where A and B are arbitrary constants. Provided $A\neq0$, this solution is singular at $r=0$.

5.2 The Dirichlet problem for a circle

As mentioned in Chapter 2, 2.3, the Dirichlet problem requires the solution of Laplace's equation

$$\nabla^2 u=0\qquad(8)$$

within a finite region R (say) with prescribed values of u on the boundary of R. The solution of this *interior* Dirichlet problem is known to be unique (see Chapter 2, 2.4). The *exterior* Dirichlet problem consists in solving (8) in the infinite region exterior to R again with prescribed values of u on the boundary of R, and can be shown in the two-dimensional case to possess a unique solution provided u remains bounded as $r\to\infty$.

We now consider the two-dimensional case where R is a circular region (see Fig. 5.1) of radius a. Then on S, the boundary of R, we have

$$u(a,\varphi)=f(\varphi),\qquad(9)$$

where f is a given function, and φ is the angular coordinate. Laplace's equation in the two-dimensional form is (from (1))

$$\frac{1}{r}\frac{\partial}{\partial r}\left(r\frac{\partial u}{\partial r}\right)+\frac{1}{r^2}\frac{\partial^2 u}{\partial\varphi^2}=0.\qquad(10)$$

Particular solutions may now be obtained by the method of separation of variables. Writing

$$u(r,\varphi)=R(r)\Phi(\varphi)\qquad(11)$$

and inserting in (10) we have

$$\frac{1}{Rr}\frac{d}{dr}\left(r\frac{dR}{dr}\right)+\frac{1}{r^2\Phi}\frac{d^2\Phi}{d\varphi^2}=0\qquad(12)$$

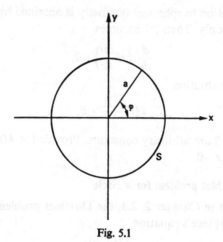

Fig. 5.1

whence putting

$$\frac{1}{\Phi}\frac{d^2\Phi}{d\varphi^2} = -\omega^2, \quad \frac{r}{R}\frac{d}{dr}\left(r\frac{dR}{dr}\right) = \omega^2 \tag{13}$$

we obtain

$$\Phi = A\cos\omega\varphi + B\sin\omega\varphi \tag{14}$$

and

$$r\frac{d}{dr}\left(r\frac{dR}{dr}\right) - \omega^2 R = 0. \tag{15}$$

Now in all physical problems $u(r, \varphi)$ must be a single-valued function such that

$$u(r, \varphi + 2\pi) = u(r, \varphi), \tag{16}$$

thus expressing the condition that on rotating through 2π we return to the same state of the system. From (8) this is possible only if

$$\omega = n, \quad n = \text{integer}. \tag{17}$$

Hence (15) becomes

$$r^2\frac{d^2R}{dr^2} + r\frac{dR}{dr} - n^2 R = 0, \tag{18}$$

which has the general solution

$$R(r) = \frac{C}{r^n} + Dr^n, \tag{19}$$

64

where C and D are arbitrary constants. Consequently using (14), (17) and (19) we have

$$u(r, \varphi) = \left(\frac{C}{r^n} + Dr^n\right) (A \cos n\varphi + B \sin n\varphi). \tag{20}$$

For the interior Dirichlet problem $(r \leqslant a)$ we must set $C = 0$ to avoid a singularity at the origin, whilst for the exterior Dirichlet problem $(r \geqslant a)$ D must be put equal to zero to ensure that $u(r, \varphi)$ remains bounded as $r \to \infty$. Hence we have

$$u_n(r, \varphi) = r^n (A_n \cos n\varphi + B_n \sin n\varphi), \quad (r \leqslant a), \tag{21}$$

$$u_n(r, \varphi) = \frac{1}{r^n} (\bar{A}_n \cos n\varphi + \bar{B}_n \sin n\varphi), \quad (r \geqslant a), \tag{22}$$

where A_n, B_n, $\bar{A}_n$, $\bar{B}_n$ are as yet arbitrary sets of constants. The linear combinations

$$u(r, \varphi) = \sum_{n=0}^{\infty} r^n (A_n \cos n\varphi + B_n \sin n\varphi), \quad \text{for } r \leqslant a, \tag{23}$$

$$u(r, \varphi) = \sum_{n=0}^{\infty} \frac{1}{r^n} (\bar{A}_n \cos n\varphi + \bar{B}_n \sin n\varphi), \quad \text{for } r \geqslant a, \tag{24}$$

are therefore harmonic functions in the two regions. We now use the boundary conditions (9) to determine the sets of coefficients in each case. From (23) we find setting $r = a$

$$f(\varphi) = \sum_{n=0}^{\infty} a^n (A_n \cos n\varphi + B_n \sin n\varphi) \tag{25}$$

whence, comparing with the Fourier expansion of $f(\varphi)$ in $-\pi \leqslant \varphi \leqslant \pi$,

$$f(\varphi) = \frac{\alpha_0}{2} + \sum_{n=1}^{\infty} (\alpha_n \cos n\varphi + \beta_n \sin n\varphi) \tag{26}$$

where

$$\alpha_0 = \frac{1}{\pi} \int_{-\pi}^{\pi} f(\lambda)d\lambda, \tag{27}$$

$$\alpha_n = \frac{1}{\pi} \int_{-\pi}^{\pi} f(\lambda) \cos n\lambda d\lambda, \quad (n = 1, 2, 3, \ldots), \tag{28}$$

$$\beta_n = \frac{1}{\pi} \int_{-\pi}^{\pi} f(\lambda) \sin n\lambda d\lambda, \quad (n = 1, 2, 3, \ldots), \tag{29}$$

we obtain

$$A_0 = \frac{\alpha_0}{2}, \quad A_n = \frac{\alpha_n}{a^n}, \quad B_n = \frac{\beta_n}{a^n}, \quad (n = 1, 2, 3, \ldots). \tag{30}$$

The solution of the interior Dirichlet problem is therefore

$$u(r, \varphi) = \frac{\alpha_0}{2} + \sum_{n=1}^{\infty} \left(\frac{r}{a}\right)^n (\alpha_n \cos n\varphi + \beta_n \sin n\varphi). \tag{31}$$

Likewise from (24) we find that

$$f(\varphi) = \sum_{n=0}^{\infty} \frac{1}{a^n} (\bar{A}_n \cos n\varphi + \bar{B}_n \sin n\varphi) \tag{32}$$

whence again using (26)–(29), we obtain

$$\bar{A}_n = \frac{\alpha_0}{2}, \quad \bar{A}_n = a^n \alpha_n, \quad \bar{B}_n = a^n \beta_n \ (n = 1, 2, 3, \ldots). \tag{33}$$

Accordingly the solution of the exterior Dirichlet problem is

$$u(r, \varphi) = \frac{\alpha_0}{2} + \sum_{n=1}^{\infty} \left(\frac{a}{r}\right)^n (\alpha_n \cos n\varphi + \beta_n \sin n\varphi). \tag{34}$$

Example 1. Suppose

$$f(\varphi) = \varphi. \tag{35}$$

Then from (27)–(29) we find

$$\alpha_0 = 0, \quad \alpha_n = 0, \quad \beta_n = \frac{1}{\pi} \int_{-\pi}^{\pi} \lambda \sin n\lambda d\lambda. \tag{36}$$

Integrating the third expression by parts we have

$$\beta_n = -\frac{2}{n}(-1)^n. \tag{37}$$

Hence from (31) and (37) the solution of the interior Dirichlet problem is

$$u(r, \varphi) = -2 \sum_{n=1}^{\infty} \left(\frac{r}{a}\right)^n (-1)^n \frac{\sin n\varphi}{n}. \tag{38}$$

Likewise from (34) and (37), the solution of the exterior Dirichlet problem is

$$u(r, \varphi) = -2 \sum_{n=1}^{\infty} \left(\frac{a}{r}\right)^n (-1)^n \frac{\sin n\varphi}{n}. \tag{39}$$

5.3 Special functions

In all the previous examples of the method of separation of variables, the solutions of boundary value problems have involved only

66

elementary functions (e.g. sines, cosines, exponentials, etc.). However, in more complicated problems, especially those concerned with the solution of the equations of mathematical physics in cylindrical and spherical regions of space, other types of functions (usually referred to as the special functions of mathematical physics) arise. Of particular importance amongst these are the Bessel and Legendre functions. The study of special functions is a subject in its own right, and in a book of this type where the emphasis is on the nature and methods of solution of partial differential equations special functions are only of secondary importance. Each particular boundary value problem when solved by the separation of variables method (if applicable), or by some other method such as a transform technique (see Chapter 8), will generate its own type of special function whose properties are almost certain to be found in the literature (see, for example, [7], [5] and [13]). However, as already mentioned, Bessel and Legendre functions are particularly important since they occur more frequently in the solution of the boundary value problems of mathematical physics than any other special function. In the remaining section of this chapter two relatively simple and typical boundary value problems which involve Bessel and Legendre functions are discussed, the few basic properties of these functions necessary for an understanding of this section (and also the relevant sections of Chapter 7 and 8) being stated without proof.

5.4 Boundary value problems involving special functions

Example 2. In cylindrical polar coordinates (r, φ, z), the heat conduction equation

$$\nabla^2 u = \frac{1}{k} \frac{\partial u}{\partial t}, \tag{40}$$

where k is a constant, takes the form (see Chapter 4, equation (15))

$$\frac{1}{r} \frac{\partial}{\partial r}\left(r \frac{\partial u}{\partial r}\right) + \frac{1}{r^2} \frac{\partial^2 u}{\partial \varphi^2} + \frac{\partial^2 u}{\partial z^2} = \frac{1}{k} \frac{\partial u}{\partial t}. \tag{41}$$

We now consider the two-dimensional form of (41) in which there is no z-dependence, and assume further that the dependent variable u is independent of the angular coordinate φ. Accordingly (41) takes the form

$$\frac{1}{r} \frac{\partial}{\partial r}\left(r \frac{\partial u}{\partial r}\right) = \frac{1}{k} \frac{\partial u}{\partial t}. \tag{42}$$

The independence of u on φ corresponds to the assumption of circular symmetry about the z-axis (see Fig. 5.2) and (42) therefore governs the radial flow of heat in a thin circular plate (r being the radial

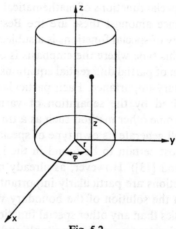

Fig. 5.2

distance from its centre). We now attempt to solve (42) subject to the boundary conditions

$$u(a, t) = 0, \quad t \geqslant 0, \tag{43}$$

$$u(r, 0) = f(r), \quad 0 \leqslant r < a, \tag{44}$$

where $f(r)$ is assumed to be a known function. Writing

$$u(r, t) = R(r)T(t), \tag{45}$$

where R and T are functions of r and t respectively, we have from (42)

$$\frac{1}{rR} \frac{d}{dr}\left(r \frac{dR}{dr}\right) = \frac{1}{kT} \frac{dT}{dt}. \tag{46}$$

Hence, in the usual way, we put

$$\frac{1}{rR} \frac{d}{dr}\left(r \frac{dR}{dr}\right) = -\lambda^2, \tag{47}$$

$$\frac{1}{kT} \frac{dT}{dt} = -\lambda^2, \tag{48}$$

where λ is an arbitrary constant. The reason for choosing $-\lambda^2$

68

rather than λ^2 in (47) and (48) will be discussed shortly. For the moment, however, (47) and (48) lead respectively to

$$r^2 \frac{d^2R}{dr^2} + r \frac{dR}{dr} + r^2\lambda^2 R = 0 \qquad (49)$$

and

$$T(t) = Ce^{-k\lambda^2 t}, \qquad (50)$$

where C is an arbitrary constant.

Equation (49) is a special case of Bessel's equation

$$x^2 \frac{d^2y}{dx^2} + x \frac{dy}{dx} + (x^2 - v^2)y = 0 \qquad (51)$$

(where v is a real constant), which has its general solution, when v is not a positive integer or zero,

$$y = AJ_v(x) + BJ_{-v}(x), \qquad (52)$$

where $J_v(x)$ and $J_{-v}(x)$ are the Bessel functions of the first kind of orders v and $-v$ respectively, and A and B are arbitrary constants of integration. These two functions may be obtained as series by the Frobenius series solution technique. However, when $v = n$ (where n is a positive integer, or zero), $J_n(x)$ and $J_{-n}(x)$ are no longer independent, but are related by the equation

$$J_{-n}(x) = (-1)^n J_n(x) \text{ (see [7])}. \qquad (53)$$

Consequently for zero or positive integral v, the Frobenius method gives only *one* solution of (51), namely $J_n(x)$ which has the form

$$J_n(x) = \sum_{r=0}^{\infty} \frac{(-1)^r}{r!(n+r)!} \left(\frac{x}{2}\right)^{n+2r} \text{(see [7])}. \qquad (54)$$

A second solution, known as Weber's solution, $Y_n(x)$ can be found by other methods, and accordingly the general solution of (51) when $v = n$ is

$$y(x) = AJ_n(x) + BY_n(x). \qquad (55)$$

We note here that, from (54),

$$J_0(x) = 1 - \frac{x^2}{(1!)^2 2^2} + \frac{x^4}{(2!)^2 2^4} - \frac{x^6}{(3!)^2 2^6} + \dots \qquad (56)$$

and

$$J_1(x) = \frac{x}{2} - \frac{x^3}{2^3 . 1!2!} + \frac{x^5}{2^5 . 2!3!} - \frac{x^7}{2^7 . 3!4!} + \dots \qquad (57)$$

69

From (56) and (57) it follows that

$$\frac{dJ_0(x)}{dx} = -J_1(x),$$ (58)

a result which will be needed shortly.

Now the solution of (49) may be obtained from (51) and (55) by writing $x = \lambda r$, $y = R$, and putting $\nu = 0$. Then

$$R(r) = AJ_0(\lambda r) + BY_0(\lambda r).$$ (59)

The graphs of the J_0 and Y_0 functions are shown in Fig. 5.3 and it is

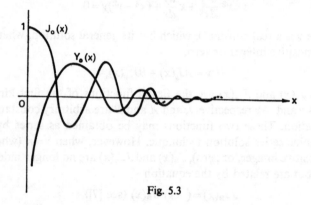

Fig. 5.3

readily seen that Y_0 is singular at the origin $r = 0$ (as indeed are all the Y_n). Hence for a non-singular solution within the circle $r = a$, we must choose $B = 0$. Consequently

$$R(r) = AJ_0(\lambda r).$$ (60)

This, together with (50) and (45), leads to

$$u(r, t) = AJ_0(\lambda r)e^{-k\lambda^2 t},$$ (61)

(the constant C in (50) being put equal to unity without loss of generality).

Now imposing the boundary condition (43) on (61) we find

$$0 = AJ_0(\lambda a)e^{-\lambda^2 kt}, \quad t \geqslant 0,$$ (62)

whence, since A cannot be set equal to zero without obtaining the trivial solution $u = 0$, we find

$$J_0(\lambda a) = 0.$$ (63)

70

This equation, which determines the eigenvalues λ of the boundary value problem defined by (51) and (43) has an infinity of solutions $\lambda_1, \lambda_2, \ldots \lambda_s, \ldots$. Writing $x = \lambda a$, the first four roots of $J_0(x) = 0$ are

$$x_1 = 2 \cdot 405, \quad x_2 = 5 \cdot 520, \quad x_3 = 8 \cdot 654, \quad x_4 = 11 \cdot 79. \tag{64}$$

At this particular point we are in a position to return to (47) and (48), and to discuss why the negative sign was chosen in front of λ^2, and indeed why λ was not put equal to zero. This last case, however, is easily disposed with since with $\lambda = 0$ (48) gives no explicit t dependence, and the boundary condition (44) cannot be imposed on the solution. In the case of $+\lambda^2$ (48) (so modified) leads to

$$T(t) = Ce^{\lambda^2 kt}, \tag{65}$$

and (49) becomes

$$r^2 \frac{d^2R}{dr^2} + r \frac{dR}{dr} - r^2\lambda^2 R = 0. \tag{66}$$

The general solution of (66) is (see, for example, [7])

$$R(r) = EI_0(\lambda r) + FK_0(\lambda r), \tag{67}$$

where I_0 and K_0 are the modified Bessel functions of the first and second kind respectively, of order zero (see Fig. 5.4), and E and F are

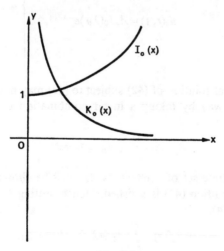

Fig. 5.4

71

arbitrary constants. The K_0 term, which is singular at $r = 0$, may be excluded by putting $F = 0$. We then have the non-singular solution

$$R(r) = EI_0(\lambda r), \tag{68}$$

whence, using (65),

$$u(r, t) = EI_0(\lambda r)e^{\lambda^2 kt}. \tag{69}$$

This is in contrast to the solution (61) obtained in the case of $-\lambda^2$. Imposing the boundary condition (43) on (69), however, now leads to

$$0 = EI_0(\lambda a)e^{\lambda^2 kt}, \quad t \geqslant 0. \tag{70}$$

But $I_0(\lambda a)$, unlike $J_0(\lambda a)$, is never zero. Hence the only way in which (70) can be satisfied is by choosing $E = 0$, and in this case, by (69), we obtain the trivial solution $u(r, t) = 0$. The choice of writing $+\lambda^2$ in (47) and (48) in place of $-\lambda^2$ must therefore be excluded.

We now return to the solution give in (61). To each of the infinity of eigenvalues $\lambda_1, \lambda_2, \lambda_3, \ldots, \lambda_r, \ldots$ obtained from (63) there corresponds a particular eigenfunction with the form of (61); these eigenfunctions are

$$\left. \begin{aligned} u_1(r, t) &= A_1 J_0(\lambda_1 r)e^{-\lambda_1^2 kt}, \\ u_2(r, t) &= A_2 J_0(\lambda_2 r)e^{-\lambda_2^2 kt}, \\ &\qquad\vdots \\ u_s(r, t) &= A_s J_0(\lambda_s r)e^{-\lambda_s^2 kt}, \\ &\qquad\vdots \end{aligned} \right\} \tag{71}$$

A more general solution of (42) subject to (43) may now be obtained in the usual way by taking a linear combination of these eigenfunctions:

$$u(r, t) = \sum_{s=1}^{\infty} A_s J_0(\lambda_s r)e^{-\lambda_s^2 kt}, \tag{72}$$

where the infinite set of coefficients A_s must be chosen so that the boundary condition (44) is satisfied. Hence putting $t = 0$ in (72) we have, using (44),

$$u(r, 0) = f(r) = \sum_{s=1}^{\infty} A_s J_0(\lambda_s r), \quad 0 \leqslant r < a. \tag{73}$$

The series (73), in which a known function is expressed as a linear combination of Bessel functions, is called a Fourier-Bessel series. To obtain the A_s in this series we multipy both sides of (73) by $rJ_0(\lambda_p r)$ and integrate from $r = 0$ to $r = a$, giving

$$\int_0^a rJ_0(\lambda_p r)f(r)dr = \int_0^a rJ_0(\lambda_p r)\left\{\sum_{s=1}^\infty A_s J_0(\lambda_s r)\right\}dr. \tag{74}$$

On the assumption that the integral and summation signs may be interchanged, this becomes

$$\int_0^a rJ_0(\lambda_p r)f(r)dr = \sum_{s=1}^\infty A_s \int_0^a rJ_0(\lambda_p r)J_0(\lambda_s r)dr. \tag{75}$$

The integral on the right of (75) may be evaluated with the help of the Lommel integrals (see, for example, [7])

$$\int_0^a rJ_n(\lambda_p r)J_n(\lambda_s r)dr = 0, \quad (p \neq s), \tag{76}$$

and

$$\int_0^a rJ_n^2(\lambda_p r)dr = \frac{a^2}{2}\left[J_n'(\lambda_p a)^2 + \left(1 - \frac{n^2}{\lambda_p^2 a^2}\right)J_n^2(\lambda_p a)\right], \tag{77}$$

where $J_n'(\lambda_p a)$ is the derivative of $J_n(x)$ at $x = \lambda_p a$.† Expanding the summation term in (74) and using (76) we find

$$\int_0^a rJ_0(\lambda_p r)f(r)dr = A_p \int_0^a rJ_0^2(\lambda_p r)dr, \quad (p = 1, 2, 3, \ldots), \tag{78}$$

whence, using (77) with $n = 0$,

$$\int_0^a rJ_0(\lambda_p r)f(r)dr = \frac{a^2 A_p}{2}[J_0'(\lambda_p a)^2 + J_0^2(\lambda_p a)]. \tag{79}$$

However, by (63), the eigenvalues λ_p are such that $J_0(\lambda_p a) = 0$. Accordingly the last term on the right-hand side of (79) is zero, and we have

$$A_p = \frac{2}{a^2[J_0'(\lambda_p a)]^2}\int_0^a rJ_0(\lambda_p r)f(r)dr, \quad (p = 1, 2, 3, \ldots). \tag{80}$$

Using the result that

$$\frac{dJ_0(x)}{dx} = -J_1(x), \quad \text{(see equation (58))}, \tag{81}$$

† We note here that (76) shows that the Bessel functions of the first kind of integral order form an orthogonal set with respect to the different eigenvalues λ_p, λ_s on the interval $0 < r < a$, with respect to a weight function $\omega(r) = r$ (see Chapter 3, 3.2).

(80) may be written finally as

$$A_p = \frac{2}{a^2 J_1^2(\lambda_p a)} \int_0^a r J_0(\lambda_p r) f(r) dr, \quad (p = 1, 2, 3, \ldots). \tag{82}$$

The solution of the boundary value problem defined by (42), (43) and (44) is therefore, by (72) and (82),

$$u(r, t) = \frac{2}{a^2} \sum_{s=1}^{\infty} \left\{ \frac{J_0(\lambda_s r)}{J_1^2(\lambda_s a)} e^{-\lambda_s^2 kt} \int_0^a r' J_0(\lambda_s r') f(r') dr' \right\}, \tag{83}$$

where r' has been written as the variable of integration to avoid confusion with the independent variable r, and where the eigenvalues λ_s are, by (63), the positive roots of $J_0(\lambda a) = 0$.

Example 3. In spherical polar coordinates (r, φ, θ), Laplace's equation

$$\nabla^2 u = 0 \tag{84}$$

takes the form (see Chapter 4, equation (18))

$$\frac{1}{r^2} \frac{\partial}{\partial r}\left(r^2 \frac{\partial u}{\partial r}\right) + \frac{1}{r^2 \sin^2 \theta} \frac{\partial^2 u}{\partial \varphi^2} + \frac{1}{r^2 \sin \theta} \frac{\partial}{\partial \theta}\left(\sin \theta \frac{\partial u}{\partial \theta}\right) = 0. \tag{85}$$

We now attempt to solve (84) in a spherical domain of radius a subject to the Dirichlet boundary conditions

$$u(a, \theta) = u_1 \ (= \text{constant}), \quad 0 \leqslant \theta < \frac{\pi}{2}, \tag{86}$$

$$u(a, \theta) = u_2 \ (= \text{constant}), \quad \frac{\pi}{2} < \theta \leqslant \pi, \tag{87}$$

where the line $\theta = 0$ is taken as the axis of symmetry (see Fig 5.5).

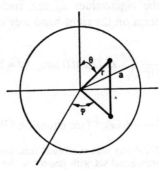

Fig. 5.5

This then is the problem of solving Laplace's equation within a sphere, the surface of the upper hemisphere having $u = u_1$, and the surface of the lower hemisphere having $u = u_2$. Since, by assumption, there is no dependence on φ, (85) may be written as

$$\frac{\partial}{\partial r}\left(r^2 \frac{\partial u}{\partial r}\right) + \frac{1}{\sin \theta} \frac{\partial}{\partial \theta}\left(\sin \theta \frac{\partial u}{\partial \theta}\right) = 0. \tag{88}$$

Writing

$$u(r, \theta) = R(r)\,\mathscr{H}\,(\theta) \tag{89}$$

and inserting into (88), we find

$$\frac{1}{R} \frac{d}{dr}\left(r^2 \frac{dR}{dr}\right) + \frac{1}{\mathscr{H} \sin \theta} \frac{d}{d\theta}\left(\sin \theta \frac{d\mathscr{H}}{d\theta}\right) = 0. \tag{90}$$

Hence

$$\frac{1}{R} \frac{d}{dr}\left(r^2 \frac{dR}{dr}\right) = \lambda, \tag{91}$$

and

$$\frac{1}{\mathscr{H} \sin \theta} \frac{d}{d\theta}\left(\sin \theta \frac{d\mathscr{H}}{d\theta}\right) = -\lambda, \tag{92}$$

where λ is a constant. Writing $\lambda = s(s+1)$, where s is a new constant, (91) is easily solved to give

$$R(r) = Ar^s + \frac{B}{r^{s+1}}, \tag{93}$$

where A and B are arbitrary constants. Equation (86) now becomes

$$\frac{1}{\sin \theta} \frac{d}{d\theta}\left(\sin \theta \frac{d\mathscr{H}}{d\theta}\right) + s(s+1)\mathscr{H} = 0. \tag{94}$$

This equation has bounded (i.e. finite) solutions in the range $0 \leqslant \theta \leqslant \pi$ only when $s = n$ (an integer or zero), and is known as Legendre's equation. These finite solutions in $0 \leqslant \theta \leqslant \pi$ are the Legendre polynomials $P_n (\cos \theta)$ which have the form (see Chapter 3, 3.2)

$$P_n(\mu) = \frac{1}{2^n n!} \frac{d^n}{d\mu^n} [(\mu^2 - 1)^n], \quad (n = 0, 1, 2, \ldots), \tag{95}$$

where $\mu = \cos \theta$.

The solution of (88) which is everywhere finite inside the sphere $r = a$ is therefore, by (89), (93) and (95),

$$u(r, \mu) = Ar^n P_n(\mu), \tag{96}$$

the constant B in (87) having been put equal to zero to exclude the singularity at the origin.

We now take a linear combination of these solutions to get

$$u(r, \mu) = \sum_{n=0}^{\infty} A_n r^n P_n(\mu), \qquad (97)$$

and attempt to choose the infinite set of coefficients A_n so that the boundary conditions (86) and (87) are satisfied. To do this we first put $r = a$ in (97) and then multiply through by $P_m(\mu)$ and integrate from $\mu = -1$ to $\mu = +1$ using the orthogonality property of the Legendre polynomials

$$\int_{-1}^{1} P_m(\mu)P_n(\mu)d\mu = 0, \quad (m \neq n). \qquad (98)$$

On the assumption that the integral and summation signs may be interchanged we are left only with the $n = m$ term which gives

$$\int_{-1}^{1} u(a, \mu)P_m(\mu)d\mu = A_m a^m \int_{-1}^{1} P_m^2(\mu)d\mu, \quad (m = 0, 1, 2, \ldots). \qquad (99)$$

Hence, using the result

$$\int_{-1}^{1} P_m^2(\mu)d\mu = \frac{2}{2m+1}, \quad (m = 0, 1, 2, \ldots), \qquad (100)$$

(Chapter 3, equation (48)), we find that (99) gives

$$A_m = \frac{2m+1}{2a^m} \int_{-1}^{1} u(a, \mu)P_m(\mu)d\mu. \qquad (101)$$

Since $\mu = \cos \theta$, the range $-1 \leqslant \mu \leqslant 0$ corresponds to $\pi \geqslant \theta \geqslant \pi/2$, and $0 \leqslant \mu \leqslant 1$ to $\pi/2 \geqslant \theta \geqslant 0$. Hence, writing (101) as

$$A_m = \frac{2m+1}{2a^m} \int_{-1}^{0} u(a, \mu)P_m(\mu)d\mu + \frac{2m+1}{2a^m} \int_{0}^{1} u(a, \mu)P_m(\mu)d\mu \qquad (102)$$

and using (86) and (87), we have finally

$$A_m = \frac{2m+1}{2a^m} u_2 \int_{-1}^{0} P_m(\mu)d\mu + \frac{2m+1}{2a^m} u_1 \int_{0}^{1} P_m(\mu)d\mu. \qquad (103)$$

These integrals may be evaluated generally by inserting the formula

76

for the Legendre polynomials given in (95). However, from (95) the first four Legendre polynomials are

$$P_0(\mu) = 1,$$
$$P_1(\mu) = \mu,$$
$$P_2(\mu) = \tfrac{1}{2}(3\mu^2 - 1),$$
$$P_3(\mu) = \tfrac{1}{2}(5\mu^3 - 3\mu).$$

(104)

Inserting these functions into (103), integrating to obtain the corresponding A_m, and finally using (97), we obtain the first few terms of the solution as

$$u(r, \theta) = \tfrac{1}{2}(u_1 + u_2) + \tfrac{3}{4}(u_2 - u_1)\left(\frac{r}{a}\right)P_1(\cos \theta) +$$

$$+ \frac{7}{16}(u_2 - u_1)\left(\frac{r}{a}\right)^3 P_3(\cos \theta) + \ldots \quad (105)$$

PROBLEMS 5

1. Show that the solution $u(r, \theta)$ of Laplace's equation $\nabla^2 u = 0$ in the semi-circular region $r < a$, $0 < \theta < \pi$, which vanishes on $\theta = 0$ and takes the constant value A on $\theta = \pi$ and on the curved boundary $r = a$, is

$$u(r, \theta) = \frac{A}{\pi}\left[\theta + 2\sum_{n=1}^{\infty}\left(\frac{r}{a}\right)^n \frac{\sin n\theta}{n}\right].$$

2. The Bessel function $J_n(x)$ of order n (n a positive or negative integer) is defined via a *generating function* by

$$e^{-x/2(t-1/t)} = \sum_{n=-\infty}^{\infty} J_n(x)t^n.$$

From this definition prove that

(i) $J_{-n}(x) = (-1)^n J_n(x)$ (see equation (53)),

(ii) $J_n(x) = \sum_{r=0}^{\infty} \frac{(-1)^r}{r!(n+r)!}\left(\frac{x}{2}\right)^{n+2r}$ (see equation (54)).

3. Verify, using the series expansion of $J_0(x)$, that

$$J_0(x) = \frac{1}{\pi}\int_0^{\pi} \cos(x \sin \varphi)d\varphi.$$

77

4. Show that the function $f(x) = 1 (0 < x < a)$ has a Fourier-Bessel expansion

$$\frac{2}{a} \sum_{n=1}^{\infty} \frac{J_0(\lambda_n x)}{\lambda_n J_1(\lambda_n a)},$$

where λ_n are the positive roots of the equation $J_0(\lambda a) = 0$.

5. Using the result of the previous question, show that the solution of the equation

$$\frac{1}{r} \frac{\partial}{\partial r} \left(r \frac{\partial u}{\partial r} \right) = \frac{1}{k} \frac{\partial u}{\partial t}$$

for $0 \leqslant r \leqslant a$, $t \geqslant 0$ (where k and a are constants) subject to the boundary conditions $u(r, 0) = T_0$ (const.), and $u(a, t) = 0$ for all t, is

$$u(r, t) = \frac{2T_0}{a} \sum_{n=1}^{\infty} \frac{J_0(\lambda_n r)}{\lambda_n J_1(\lambda_n a)} e^{-k\lambda_n^2 t},$$

where λ_n are the positive roots of $J_0(\lambda a) = 0$.

CHAPTER 6

Continuous Eigenvalues and Fourier Integrals

6.1 Introduction

In each of the previous examples of the method of separation of variables, the eigenvalues of the resulting Sturm-Liouville equation formed a discrete infinite set. To each of these eigenvalues there corresponded an eigenfunction, and a linear combination of these eigenfunctions served as the solution of the partial differential equation. However, not all boundary value problems lead to just this state of affairs. Consider, for example, the problem of solving the heat conduction equation

$$\frac{\partial^2 u}{\partial x^2} = \frac{1}{k}\frac{\partial u}{\partial t} \tag{1}$$

subject to the boundary conditions

$$u(0, t) = 0, \quad t > 0, \tag{2}$$

$$u(x, 0) = f(x), \quad x \geqslant 0, \tag{3}$$

the solution $u(x, t)$ to be bounded (that is, finite) in the semi-infinite space interval $0 < x < \infty$. (This problem differs from the one posed in Chapter 4, Example 1, where a solution was required on the *finite* interval $0 \leqslant x \leqslant l$. Here the solution is required for all positive x.)

Now writing

$$u(x, t) = X(x)T(t) \tag{4}$$

we find, in the usual way,

$$\frac{d^2 X}{dx^2} + \lambda^2 X = 0, \tag{5}$$

$$\frac{dT}{dt} + \lambda^2 kT = 0, \tag{6}$$

79

where the sign of the separation constant (λ^2) has been so chosen that bounded solutions for both X and T arise. These solutions are

$$X(x) = A \cos \lambda x + B \sin \lambda x, \tag{7}$$

$$T(t) = Ce^{-\lambda^2 kt}, \tag{8}$$

where A, B and C are arbitrary constants of integration. Hence

$$u(x, t) = (A \cos \lambda x + B \sin \lambda x)e^{-\lambda^2 kt}, \tag{9}$$

the constant C having been put equal to unity without loss of generality. Imposing the first boundary condition (2), we find

$$A = 0, \tag{10}$$

and hence

$$u(x, t) = Be^{-\lambda^2 kt} \sin \lambda x. \tag{11}$$

Unlike the related problem (Chapter 4, Example 1) where the boundary condition at $x = l$ gave the permitted values of λ (the eigenvalues), we now have no such condition. Accordingly all real positive values of λ are permissible, and instead of a sum over an infinite discrete set of λ values we now have to take an integral over λ to obtain the general solution

$$u(x, t) = \int_{\lambda=0}^{\infty} B(\lambda)e^{-\lambda^2 kt} \sin \lambda x d\lambda, \tag{12}$$

where $B(\lambda)$ is an arbitrary function of λ.

Imposing the boundary condition (3) on (12) we have

$$f(x) = \int_{0}^{\infty} B(\lambda) \sin \lambda x d\lambda, \tag{13}$$

from which we must determine $B(\lambda)$. Equation (13) is, in fact, an integral equation since the unknown function occurs in the integrand. This particular type of integral equation can be readily solved with the help of the Fourier integral formula, which we discuss in the next section.

6.2 The Fourier integral

We have seen earlier that subject to certain conditions (the Dirichlet conditions) we may expand a function $f(x)$ in the finite interval $-l < x < l$ as

$$f(x) = \frac{a_0}{2} + \sum_{r=1}^{\infty} \left(a_r \cos \frac{\pi r x}{l} + b_r \sin \frac{\pi r x}{l} \right), \tag{14}$$

where

$$a_r = \frac{1}{l} \int_{-l}^{l} f(\xi) \cos \frac{\pi r \xi}{l} \, d\xi, \quad (r = 0, 1, 2, 3, \ldots), \tag{15}$$

and

$$b_r = \frac{1}{l} \int_{-l}^{l} f(\xi) \sin \frac{\pi r \xi}{l} \, d\xi, \quad (r = 1, 2, 3, \ldots), \tag{16}$$

ξ being the variable of integration. Combining (14), (15) and (16) we have

$$f(x) = \frac{1}{2l} \int_{-l}^{l} f(\xi) d\xi + \sum_{r=1}^{\infty} \left\{ \frac{1}{l} \left[\int_{-l}^{l} f(\xi) \cos \frac{\pi r \xi}{l} \, d\xi \right] \cos \frac{\pi r x}{l} + \right.$$
$$\left. + \frac{1}{l} \left[\int_{-l}^{l} f(\xi) \sin \frac{\pi r \xi}{l} \, d\xi \right] \sin \frac{\pi r x}{l} \right\}. \tag{17}$$

The representation (17) is valid for $-l < x < l$, but since the right-hand side is a function of period $2l$ it cannot represent $f(x)$ *outside* this range unless $f(x)$ also is a periodic function of period $2l$. We now attempt to find a representation of a non-periodic function in the range $-\infty < x < \infty$ by letting $l \to \infty$. Then, provided $f(x)$ is such that the integral

$$\int_{-\infty}^{\infty} f(x) dx \tag{18}$$

exists (i.e. is finite), the first term of (17) is zero in the limit. In the remaining terms on the right-hand side of (17) as $l \to \infty$, $\frac{1}{l}$ becomes infinitesimally small. The sum of these terms, however, does not necessarily vanish since, as $l \to \infty$, the number of terms increases.

Writing $\triangle \lambda = \pi/l$, we have (from (17))

$$f(x) = \frac{1}{\pi} \sum_{r=1}^{\infty} \triangle \lambda \left\{ \int_{-l}^{l} f(\xi) \cos [r \triangle \lambda (\xi - x)] d\xi \right\}. \tag{19}$$

As $l \to \infty$, $\triangle \lambda \to 0$, and the sum in (19) can be replaced (under suitable conditions which we will not enter into here) by an integral to give

$$f(x) = \frac{1}{\pi} \int_{0}^{\infty} d\lambda \int_{-\infty}^{\infty} f(\xi) \cos \lambda(\xi - x) d\xi, \quad (-\infty < x < \infty). \tag{20}$$

This is the Fourier integral formula, and may be written in the form

$$f(x) = \int_{0}^{\infty} [A(\lambda) \cos \lambda x + B(\lambda) \sin \lambda x] d\lambda, \tag{21}$$

where

$$A(\lambda) = \frac{1}{\pi} \int_{-\infty}^{\infty} f(\xi) \cos \lambda\xi d\xi \qquad (22)$$

and

$$B(\lambda) = \frac{1}{\pi} \int_{-\infty}^{\infty} f(\xi) \sin \lambda\xi d\xi. \qquad (23)$$

Two special cases of the Fourier integral exist.

If $f(x)$ is an even function in $-\infty < x < \infty$, then $f(\xi) = f(-\xi)$, and (22) and (23) become respectively

$$A(\lambda) = \frac{2}{\pi} \int_{0}^{\infty} f(\xi) \cos \lambda\xi d\xi \qquad (24)$$

and

$$B(\lambda) = 0. \qquad (25)$$

Hence inserting (24) and (25) in (21) we obtain the Fourier cosine integral

$$f(x) = \frac{2}{\pi} \int_{0}^{\infty} \cos \lambda x \left[\int_{0}^{\infty} f(\xi) \cos \lambda\xi d\xi \right] d\lambda. \qquad (26)$$

Similarly if $f(x)$ is an odd function in $-\infty < x < \infty$, then $f(\xi) = -f(-\xi)$, and (22) and (23) become

$$A(\lambda) = 0, \qquad (27)$$

and

$$B(\lambda) = \frac{2}{\pi} \int_{0}^{\infty} f(\xi) \sin \lambda\xi d\xi \qquad (28)$$

respectively. With these results we obtain the Fourier sine integral

$$f(x) = \frac{2}{\pi} \int_{0}^{\infty} \sin \lambda x \left[\int_{0}^{\infty} f(\xi) \sin \lambda\xi d\xi \right] d\lambda. \qquad (29)$$

As with Fourier series expansions, so it may be proved that (for functions for which a Fourier integral representation is possible) the integral representation converges to $f(x)$ at all points where $f(x)$ is continuous. Similarly, at a point $(x_0$, say) of finite discontinuity of $f(x)$ the Fourier integral converges to the mean of the right-hand and left-hand limits of $f(x)$ as $x \to x_0$. These results are demonstrated more fully in the following example.

Example 1. Using the cosine integral (26), and writing

$$f(x) = e^{-kx}, \tag{30}$$

where k is a positive constant, we have

$$e^{-kx} = \frac{2}{\pi} \int_0^\infty \cos \lambda x \left[\int_0^\infty e^{-k\xi} \cos \lambda \xi d\xi \right] d\lambda. \tag{31}$$

Evaluating the integral over ξ by parts, we readily obtain the integral representation

$$e^{-kx} = \frac{2}{\pi} \int_0^\infty \frac{k}{k^2 + \lambda^2} \cos \lambda x d\lambda \tag{32}$$

which is valid for $x > 0$, $k > 0$. Now, since we have used the cosine integral, the function so represented must be an *even* function. Accordingly for negative x the function represented by (32) is the *even* extension of e^{-kx} (see Fig. 6.1). At $x = 0$ we see that the function

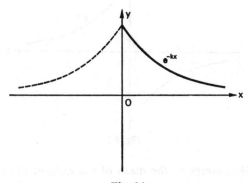

Fig. 6.1

is continuous, and consequently the Fourier cosine integral represents the function at this point also. Accordingly (32) is valid for $x \geqslant 0$, $k > 0$. This may be easily verified by checking that the right- and left-hand sides of (32) are in fact equal when $x = 0$. In this way we find

$$e^0 = \frac{2}{\pi} \int_0^\infty \frac{k}{k^2 + \lambda^2} d\lambda = \frac{2}{\pi} \left(\tan^{-1} \frac{\lambda}{k} \right)_0^\infty = 1, \tag{33}$$

which is true.

On the other hand, using the sine integral (29), we find (for $k>0$)

$$e^{-kx} = \frac{2}{\pi} \int_0^\infty \sin \lambda x \left(\int_0^\infty e^{-k\xi} \sin \lambda\xi d\xi \right) d\lambda \qquad (34)$$

$$= \frac{2}{\pi} \int_0^\infty \frac{\lambda}{k^2+\lambda^2} \sin \lambda x d\lambda, \quad (x>0, \ k>0). \qquad (35)$$

Unlike the cosine representation (32), (35) is not valid at $x=0$. For inserting $x=0$ on to the right-hand side of (35) we obtain zero, whilst at $x=0$ the left-hand side of (32) is equal to unity. The reason for this is clear when we recall that since the Fourier sine integral has been used, the function represented by (35) for negative x is the *odd* extension of e^{-kx} (see Fig. 6.2). At $x=0$ there is a finite discontinuity, and accordingly by the results already stated the Fourier

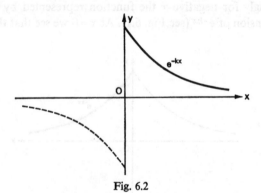

Fig. 6.2

integral will converge to the mean of the right-hand and left-hand limits of $f(x)$ as $x \to 0$. In this case this mean-value is

$$\tfrac{1}{2}[1+(-1)] = 0, \qquad (36)$$

which, as already verified, is the value of the right-hand side of (35) when $x=0$.

6.3 Application of Fourier integrals to boundary value problems

Returning to (13), we now see (by (28) and (29)) that

$$B(\lambda) = \frac{2}{\pi} \int_0^\infty f(\xi) \sin \lambda\xi d\xi, \quad (\lambda>0). \qquad (37)$$

Hence, by (12), the required solution of the boundary value problem defined by (1), (2) and (3) is

$$u(x, t) = \frac{2}{\pi} \int_0^\infty e^{-\lambda^2 kt} \sin \lambda x \left[\int_0^\infty f(\xi) \sin \lambda \xi \, d\xi \right] d\lambda. \qquad (38)$$

Example 2. To obtain a bounded solution of Laplace's equation

$$\nabla^2 u = 0 \qquad (39)$$

in the semi-infinite region defined by $x \geqslant 0$, $0 \leqslant y \leqslant 1$ (see Fig. 6.3) subject to the boundary conditions

$$\left(\frac{\partial u}{\partial x} \right)_{x=0} = 0, \quad \left(\frac{\partial u}{\partial y} \right)_{y=0} = 0, \qquad (40)$$

and

$$u(x, 1) = f(x), \qquad (41)$$

where $f(x)$ is assumed to be known.

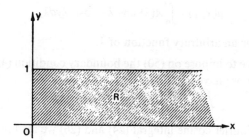

Fig. 6.3

Writing in the usual way

$$u(x, y) = X(x)\, Y(y) \qquad (42)$$

(39) becomes

$$\frac{1}{X} \frac{d^2 X}{dx^2} + \frac{1}{Y} \frac{d^2 Y}{dy^2} = 0, \qquad (43)$$

whence

$$\frac{d^2 X}{dx^2} = -\lambda^2 X, \quad \frac{d^2 Y}{dy^2} = \lambda^2 Y, \qquad (44)$$

where λ^2 is a separation constant (its sign being so chosen that the

85

resulting solution $X(x)$ when subjected to the given boundary conditions is bounded). The solutions of (44) are

$$X(x) = A \cos \lambda x + B \sin \lambda x, \tag{45}$$

$$Y(y) = C \cosh \lambda y + D \sinh \lambda y, \tag{46}$$

whence

$$u(x, y) = (A \cos \lambda x + B \sin \lambda x)(C \cosh \lambda y + D \sinh \lambda y).$$

Now imposing the first of the two boundary conditions of (40) we find that

$$B = 0, \tag{47}$$

whilst the second boundary condition of (40) leads to

$$D = 0. \tag{48}$$

Accordingly the solution now takes the form

$$u(x, y) = A \cos \lambda x \cosh \lambda y. \tag{49}$$

All real positive values of λ are permissible, and hence the general solution of (39) subject to (40) becomes

$$u(x, y) = \int_0^\infty A(\lambda) \cos \lambda x \cosh \lambda y\, d\lambda, \tag{50}$$

where $A(\lambda)$ is an arbitrary function of λ.

We now have to impose on (50) the boundary condition (41). Putting $y = 1$ in (50), we have

$$f(x) = \int_0^\infty A(\lambda) \cos \lambda x \cosh \lambda\, d\lambda. \tag{51}$$

Using the Fourier cosine integral (24) and (26) we readily see that

$$A(\lambda) \cosh \lambda = \frac{2}{\pi} \int_0^\infty f(\xi) \cos \lambda\xi\, d\xi. \tag{52}$$

Hence

$$A(\lambda) = \frac{2}{\pi \cosh \lambda} \int_0^\infty f(\xi) \cos \xi\lambda\, d\xi \tag{53}$$

and finally

$$u(x, y) = \frac{2}{\pi} \int_0^\infty \frac{\cos \lambda x \cosh \lambda y}{\cosh \lambda} \left[\int_0^\infty f(\xi) \cos \xi\lambda\, d\xi \right] d\lambda. \tag{54}$$

As a specific example, we take

$$f(x) = \begin{cases} 1, & 0 \leqslant x \leqslant 1, \tag{55} \\ 0, & x > 1. \tag{56} \end{cases}$$

Hence

$$\int_0^\infty f(\xi) \cos \xi\lambda d\xi = \int_0^1 1 \cos \xi\lambda d\xi + \int_1^\infty 0 \cos \xi\lambda d\xi \qquad (52)$$

$$= \left(\frac{\sin \lambda\xi}{\lambda}\right)_0^1 = \frac{\sin \lambda}{\lambda} \qquad (58)$$

and, by (54), therefore

$$u(x, y) = \frac{2}{\pi} \int_0^\infty \frac{\cos \lambda x \cosh \lambda y \sin \lambda}{\lambda \cosh \lambda} d\lambda. \qquad (59)$$

The two boundary value problems discussed in this section and 6.2 are typical of the type of problem which may be solved by Fourier integral methods. The common and basic feature of these problems is that a solution is required over an *infinite* space domain, and that this feature in turn leads to a *continuous* set of eigenvalues. An integral over the corresponding continuous set of eigenfunctions then serves as a solution of the problem, the Fourier integral being used to determine the unknown coefficients (functions) $A(\lambda)$ and $B(\lambda)$. In the following chapter a more unified approach to integral methods (including the Fourier integral) will be discussed, these methods usually being referred to as integral transform methods.

PROBLEMS 6

1. Given that

$$f(x) = \begin{cases} 1, & 0 < x < 1, \\ 0, & x \geqslant 1, \end{cases}$$

show, using the Fourier cosine integral, that in $0 < x < 1$

$$1 = \frac{2}{\pi} \int_0^\infty \frac{\sin \lambda \cos \lambda x}{\lambda} d\lambda.$$

2. Show that the Fourier cosine representation of the function $f(x)$ defined by

$$f(x) = \begin{cases} x, & 0 < x < 1, \\ 2 - x, & 1 < x < 2, \\ 0, & x > 2, \end{cases}$$

is

$$\frac{2}{\pi} \int_0^\infty \frac{2 \cos \lambda - \cos 2\lambda - 1}{\lambda^2} \cos \lambda x d\lambda.$$

3. Show that the solution of Laplace's equation $\nabla^2 u = 0$ in the region $0 < y < a$, $x > 0$ satisfying $u(x, 0) = f(x)$ and $u(x, a) = 0$ (where $f(x)$ is a given function, and a is a constant) is

$$u(x, y) = \frac{1}{\pi} \int_0^\infty \left[\int_{-\infty}^\infty \frac{\sinh \lambda(a - y)}{\sinh \lambda a} f(\xi) \cos \lambda(\xi - x) d\xi \right] d\lambda.$$

4. Show that the bounded solution $u(x, y)$ of Laplace's equation $\nabla^2 u = 0$ in the quadrant $x > 0$, $y > 0$, which satisfies $u(0, y) = 0$ (for $y > 0$), and $u(x, 0) = f(x)$ (for $x > 0$) is

$$u(x, y) = \frac{2}{\pi} \int_0^\infty \int_0^\infty e^{-\lambda y} f(\xi) \sin \lambda x \sin \lambda \xi d\xi d\lambda.$$

By integrating with respect to λ, transform this solution into the form

$$u(x, y) = \frac{y}{\pi} \int_0^\infty f(\xi) \left[\frac{1}{y^2 + (\xi - x)^2} - \frac{1}{y^2 + (\xi + x)^2} \right] d\xi.$$

5. Show that the solution $u(r, t)$ of the heat conduction equation

$$\frac{1}{r} \frac{\partial}{\partial r} \left(r \frac{\partial u}{\partial r} \right) = \frac{1}{k} \frac{\partial u}{\partial t} \quad (0 < r < \infty, \ t > 0)$$

which remains finite at $r = 0$, and for which $u(r, 0) = f(r)$ (where $f(r)$ is assumed known) may be written as

$$u(r, t) = \int_0^\infty A(\lambda) e^{-\lambda^2 k t} J_0(\lambda r) d\lambda,$$

where

$$f(r) = \int_0^\infty A(\lambda) J_0(\lambda r) d\lambda.$$

CHAPTER 7

The Laplace Transform

7.1 Integral transforms

In the previous chapter we obtained through the Fourier cosine integral the result that under certain conditions a function $f(x)$ may be represented in $0 < x < \infty$ in the form

$$f(x) = \int_0^\infty A(\lambda) \cos \lambda x d\lambda \qquad (1)$$

where

$$A(\lambda) = \frac{2}{\pi} \int_0^\infty f(x) \cos \lambda x dx. \qquad (2)$$

Changing the notation slightly by writing p for λ and $\bar{f}(p)$ for $\frac{\pi}{2} A(\lambda)$, we have

$$\bar{f}(p) = \int_0^\infty f(x) \cos px dx \qquad (3)$$

and

$$f(x) = \frac{2}{\pi} \int_0^\infty \bar{f}(p) \cos px dp. \qquad (4)$$

Equation (3) defines the Fourier cosine transform, $\bar{f}(p)$, of $f(x)$, whilst (4) defines the inverse Fourier cosine transform, $f(x)$, of $\bar{f}(p)$. We may therefore go from $f(x) \Leftrightarrow \bar{f}(p)$ by this pair of equations.

Similarly the Fourier sine integral leads to the pair of equations

$$\bar{f}(p) = \int_0^\infty f(x) \sin px dx \qquad (5)$$

and

$$f(x) = \frac{2}{\pi} \int_0^\infty \bar{f}(p) \sin px dp \qquad (6)$$

which define, respectively, the Fourier sine transform, $\bar{f}(p)$ of $f(x)$, and its inverse.

Both (3) and (5) are particular examples of the equation

$$\bar{f}(p) = \int_a^b f(x)K(p, x)dx \tag{7}$$

which defines $\bar{f}(p)$ to be an *integral transform* of the function $f(x)$ with respect to the function $K(p, x)$, a and b being real constants. The function $K(p, x)$ is called the *kernel* of the transform. By choosing particular forms for $K(p, x)$, and assigning values to a and b, different transforms may be set up. For example, the Fourier cosine and sine transforms are obtained from (7) by choosing $a = 0$, $b = \infty$, and letting $K(p, x)$ be, respectively, $\cos px$ and $\sin px$.

Integral transforms are of importance in the solution of certain problems where it is often easier to transform to the variable p, solve for $\bar{f}(p)$, and then use the inverse transform to recover the solution of the original problem. This, as we shall see in later sections of this chapter, and in Chapter 8, is a particularly useful approach to the solution of ordinary and partial differential equations. However, the nature of the inverse transform depends very much on the particular type of transform in use. For the Fourier cosine and sine transforms, the inversion formulae (4) and (6) are relatively simple (although when $\bar{f}(p)$ is given the integrals themselves may be difficult to evaluate). For other transforms, however, the inversion formulae may involve a technique of complex variable theory called *contour integration*. One transform for which this is the case is the Laplace transform defined by putting $a = 0$, $b = \infty$, and $K(p, x) = e^{-px}$ in (7) so that

$$\bar{f}(p) = \int_0^\infty f(x)e^{-px}dx. \tag{8}$$

This particular transform is of such importance to physicists and engineers alike that we shall devote most of this chapter to a discussion of its properties and uses. In Chapter 8, the Laplace transform, and to a lesser extent the Fourier cosine and sine transforms, will be used to solve certain types of partial differential equations.

Finally, we note that the operation of taking an integral transform (in the sense of (7)) possesses the linearity property. For suppose $I\{ \ \}$ denotes taking an integral transform of whatever function occurs inside the curly bracket; then

$$I\{f(x)\} = \bar{f}(p) = \int_a^b f(x)K(p, x)dx. \tag{9}$$

It is readily seen that if α, β are arbitrary (real) constants, and $g(x)$ is an arbitrary function for which the transform exists, we have

$$I\{\alpha f(x)\} = \alpha I\{f(x)\}, \tag{10}$$

and

$$I\{\alpha f(x) + \beta g(x)\} = \int_a^b [\alpha f(x) + \beta g(x)]K(p, x)dx \tag{11}$$

$$= \alpha I\{f(x)\} + \beta I\{g(x)\}. \tag{12}$$

Equations (10) and (12) show that $I\{\ \}$ is a linear operator.

We now introduce the inverse operator $I^{-1}\{\ \}$ which is such that if

$$I\{f(x)\} = \bar{f}(p) \tag{13}$$

then

$$f(x) = I^{-1}\{\bar{f}(p)\}. \tag{14}$$

In other words, given $\bar{f}(p)$, $I^{-1}\{\bar{f}(p)\}$ is the function from which $\bar{f}(p)$ may be derived by taking its integral transform. Clearly from (13) and (14)

$$I\{I^{-1}[\bar{f}(p)]\} = \bar{f}(p), \tag{15}$$

$$I^{-1}\{I[f(x)]\} = f(x), \tag{16}$$

whence the operators I and I^{-1} commute and have a 'product' II^{-1} $(=I^{-1}I) = 1$ (the unit operator). In virtue of the linearity of I we now have

$$I^{-1}\{\alpha f(x)\} = \alpha I^{-1}\{f(x)\}, \tag{17}$$

and

$$I^{-1}\{\alpha f(x) + \beta g(x)\} = \alpha I^{-1}\{f(x)\} + \beta I^{-1}\{g(x)\}, \tag{18}$$

where again α and β are arbitrary constants, and $f(x)$ and $g(x)$ are arbitrary functions for which the transforms exist.

7.2 The Laplace transform

Let $f(x)$ be defined for $x > 0$. Then the Laplace transform of $f(x)$ is defined by (see (8))

$$L\{f(x)\} = \bar{f}(p) = \int_0^\infty f(x)e^{-px}dx \tag{19}$$

where p is assumed here to be real (although, in general, it may be complex), and $L\{\ \}$ has been used in place of $I\{\ \}$ to denote, specifically, the Laplace transform.

The Laplace transform can be proved to exist for any function which is integrable over every finite interval $0 < x < l$ (where l is any positive number), and which, as $x \to \infty$, does not grow faster than some exponential function. This last condition is more precisely stated by writing

$$|f(x)| < Me^{\alpha x} \quad \text{as} \quad x \to \infty, \tag{20}$$

where M and α are some real positive numbers. Functions satisfying (20) are said to be of exponential order (e.g. x^k, $\sin kx$, $\cos kx$, e^{kx}, $J_0(x)$ for any k). However, the function e^{x^2} is not of exponential order since $e^{x^2}e^{-\alpha x}$ does not remain less than some number M as $x \to \infty$, but can be made indefinitely large by choosing sufficiently large values of x. Consequently the Laplace transform of this function does not exist.

We now derive the transforms of some elementary functions.

(i) If $f(x) = 1$, then

$$L\{1\} = \bar{f}(p) = \int_0^\infty e^{-px} dx = \frac{1}{p}, \tag{21}$$

where, for the existence of the integral, $p > 0$.

(ii) If $f(x) = x^n$ (where n is an integer $\geqslant 1$), then

$$L\{x^n\} = \bar{f}(p) = \int_0^\infty x^n e^{-px} dx. \tag{22}$$

Writing $px = u$, (22) becomes (for $p > 0$)

$$L\{x^n\} = \frac{1}{p^{n+1}} \int_0^\infty u^n e^{-u} du, \tag{23}$$

which by integrating by parts, finally gives

$$L\{x^n\} = \frac{n!}{p^{n+1}}, \quad n = 1, 2, 3, \ldots. \tag{24}$$

(iii) If $f(x) = e^{ax}$, where a is a real constant, then

$$L\{e^{ax}\} = \bar{f}(p) = \int_0^\infty e^{ax} e^{-px} dx \tag{25}$$

$$= \frac{1}{p-a} \quad \text{for} \quad p > a. \tag{26}$$

(iv) If $f(x) = \sin ax$, where a is a real constant, then

$$L\{\sin ax\} = \bar{f}(p) = \int_0^\infty e^{-px} \sin ax\, dx = \frac{a}{p^2 + a^2} \tag{27}$$

provided $p > 0$.

(v) If $f(x) = \cos ax$, where a is a real constant, then

$$L\{\cos ax\} = \bar{f}(p) = \int_0^\infty e^{-px} \cos ax\, dx = \frac{p}{p^2 + a^2} \qquad (28)$$

provided $p > 0$.

(vi) If $f(x) = \sinh ax$, where a is a real constant, then

$$L\{\sinh ax\} = \bar{f}(p) = \int_0^\infty e^{-px} \sinh ax\, dx = \frac{a}{p^2 - a^2} \qquad (29)$$

provided $p > |a|$.

Transforms of many other functions may be obtained in a similar way, but in some cases it is convenient to use the linearity property of the operator $L\{\ \}$. For example, instead of performing an integration in (vi) we could write

$$L\{\sinh ax\} = L\left\{\frac{e^{ax} - e^{-ax}}{2}\right\} = \tfrac{1}{2}L\{e^{ax}\} - \tfrac{1}{2}L\{e^{-ax}\} \qquad (30)$$

$$= \tfrac{1}{2}\frac{1}{p-a} - \tfrac{1}{2}\frac{1}{p+a}, \quad \text{using (26)}, \qquad (31)$$

$$= \frac{a}{p^2 - a^2}, \quad \text{for } p > |a|. \qquad (32)$$

Similarly, if $f(x) = J_0(x)$ (the Bessel function of zero order) then, using the series form (Chapter 5, (54))

$$L\{J_0(x)\} = L\left\{1 - \frac{x^2}{2^2} + \frac{x^4}{2^2.4^2} - \frac{x^6}{2^2.4^2.6^2} + \dots\right\} \qquad (33)$$

$$= \frac{1}{p} - \frac{1}{2^2} \cdot \frac{2!}{p^3} + \frac{1}{2^2.4^2}\frac{4!}{p^5} - \frac{1}{2^2.4^2.6^2} \cdot \frac{6!}{p^7} + \dots \qquad (34)$$

$$\text{(using (12) and (24))}$$

$$= \frac{1}{p}\left[1 - \frac{1}{2}\left(\frac{1}{p^2}\right) + \frac{1.3}{2.4}\left(\frac{1}{p^4}\right) - \frac{1.3.5}{2.4.6}\left(\frac{1}{p^6}\right) + \dots\right] \qquad (35)$$

$$= \frac{1}{p}\left(1 + \frac{1}{p^2}\right)^{-1/2} \quad \text{(using the binomial theorem)} \qquad (36)$$

$$= \frac{1}{\sqrt{(1 + p^2)}}. \qquad (37)$$

We may also prove that

$$L\{J_0(ax)\} = \frac{1}{\sqrt{(a^2 + p^2)}}, \quad \text{where } a \text{ is a real constant.} \tag{38}$$

We now state and prove two useful theorems.

Theorem 1. (Shift theorem)
If

$$L\{f(x)\} = \bar{f}(p) \tag{39}$$

then

$$L\{e^{-ax}f(x)\} = \bar{f}(p + a), \tag{40}$$

where a is a real constant.

The proof of this theorem is easily obtained since

$$L\{e^{-ax}f(x)\} = \int_0^\infty e^{-px} e^{-ax} f(x) dx \tag{41}$$

$$= \int_0^\infty e^{-(p+a)x} f(x) dx = \bar{f}(p + a). \tag{42}$$

Example 1. Since, by (24),

$$L\{x^n\} = \frac{n!}{p^{n+1}}, \quad n = 1, 2, 3, \ldots, \tag{43}$$

it follows, using (40), that

$$L\{x^n e^{-ax}\} = \frac{n!}{(p + a)^{n+1}}, \quad \text{for } p > -a. \tag{44}$$

Theorem 2.
If

$$L\{f(x)\} = \bar{f}(p) \tag{45}$$

then

$$L\{x^n f(x)\} = (-1)^n \frac{d^n}{dp^n} \bar{f}(p), \quad n = 1, 2, 3, \ldots \tag{46}$$

Now since

$$\bar{f}(p) = \int_0^\infty e^{-px} f(x) dx \tag{47}$$

we have

$$\frac{d}{dp}\bar{f}(p) = \frac{d}{dp}\int_0^\infty e^{-px}f(x)dx = \int_0^\infty \frac{\partial}{\partial p}[e^{-px}f(x)]dx \qquad (48)$$

$$= \int_0^\infty -xe^{-px}f(x)dx = -\int_0^\infty e^{-px}[xf(x)]dx \qquad (49)$$

$$= -L\{xf(x)\}. \qquad (50)$$

This proves (46) for the case $n = 1$. The proof may be readily extended to other values of n values by induction.

Example 2. Since (by (28))

$$L\{\cos ax\} = \frac{p}{p^2 + a^2} \qquad (51)$$

we have, using (46),

$$L\{x \cos ax\} = -\frac{d}{dp}\left(\frac{p}{p^2 + a^2}\right) = \frac{p^2 - a^2}{(p^2 + a^2)^2}. \qquad (52)$$

7.3 Inverse Laplace transforms

The inverse Laplace transform defined by

$$f(x) = L^{-1}\{\bar{f}(p)\} \quad \text{(compare with (14))} \qquad (53)$$

may be obtained by a number of different methods, the most direct of which is via a set of standard transforms. For example, since (by (21))

$$L\{1\} = \frac{1}{p}, \qquad (54)$$

we have

$$L^{-1}\left\{\frac{1}{p}\right\} = 1. \qquad (55)$$

Similarly, since (by (28))

$$L\{\cos ax\} = \frac{p}{p^2 + a^2}, \qquad (56)$$

we have

$$L^{-1}\left\{\frac{p}{p^2 + a^2}\right\} = \cos ax. \qquad (57)$$

When $\bar{f}(p)$ is a rational function of p, but is not immediately

95

recognisable as a standard type, it may very often be expressed, using partial fractions, as the sum of a number of terms each of which may be inverted at sight. The following examples illustrate this method.

Example 3. To find

$$L^{-1}\left\{\frac{1}{(p+a)(p+b)}\right\}, \tag{58}$$

where a and b are real (unequal) constants, we first write

$$\frac{1}{(p+a)(p+b)} = \frac{A}{p+a} + \frac{B}{p+b}. \tag{59}$$

Comparing coefficients of powers of p on each side of (59) we find

$$A = -B = \frac{1}{b-a}. \tag{60}$$

Consequently, using the linearity property of the operator L^{-1}, we have

$$L^{-1}\left\{\frac{1}{(p+a)(p+b)}\right\} = \frac{1}{b-a}L^{-1}\left\{\frac{1}{p+a}\right\} - \frac{1}{b-a}L^{-1}\left\{\frac{1}{p+b}\right\} \tag{61}$$

$$= \frac{1}{b-a}(e^{-ax} - e^{-bx}), \quad \text{(using 26))}. \tag{62}$$

Example 4. To find

$$L^{-1}\left\{\frac{p}{(p^2+a^2)(p^2+b^2)}\right\}, \tag{63}$$

where a and b are real (unequal) constants, we write

$$\frac{p}{(p^2+a^2)(p^2+b^2)} = \frac{1}{b^2-a^2}\left(\frac{p}{p^2+a^2} - \frac{p}{p^2+b^2}\right). \tag{64}$$

Consequently

$$L^{-1}\left\{\frac{p}{(p^2+a^2)(p^2+b^2)}\right\} = \frac{1}{b^2-a^2}(\cos ax - \cos bx). \tag{65}$$

When the inverse transform of a function cannot be obtained directly from a set of standard transforms or by the method of partial fractions, some other approach must be used. To this end, we now state (without proof) the following theorem.

96

Theorem 3. (Convolution theorem)

Let $f(x)$ and $g(x)$ be two arbitrary functions each possessing a Laplace transform so that

$$\bar{f}(p) = L\{f(x)\}, \tag{66}$$

$$\bar{g}(p) = L\{g(x)\}. \tag{67}$$

Then, under certain conditions, it can be proved that

$$L\left\{\int_0^x f(x-u)g(u)du\right\} = \bar{f}(p)\bar{g}(p), \tag{68}$$

whence

$$\int_0^x f(x-u)g(u)du = L^{-1}\{\bar{f}(p)\bar{g}(p)\}. \tag{69}$$

Example 5. To show how the convolution theorem can be used to evaluate inverse transforms, consider the problem of finding

$$L^{-1}\left\{\frac{1}{p^2(p+1)^2}\right\}. \tag{70}$$

Now by (24) and (44) respectively, we have

$$L^{-1}\left\{\frac{1}{p^2}\right\} = x, \quad L^{-1}\left\{\frac{1}{(p+1)^2}\right\} = xe^{-x}. \tag{71}$$

Consequently letting $f(x) \equiv x$, $g(x) \equiv xe^{-x}$, and using (69) and (71)

$$L^{-1}\left\{\frac{1}{p^2(p+1)^2}\right\} = L^{-1}\left\{\frac{1}{p^2}\frac{1}{(p+1)^2}\right\} = L^{-1}\{\bar{f}(p)\bar{g}(p)\} \tag{72}$$

$$= \int_0^x (x-u)ue^{-u}du \tag{73}$$

$$= (x+2)e^{-x} + x - 2. \tag{74}$$

If none of the methods described here leads to the inverse of a given function, it is probably necessary to resort to the general inversion formula of the Laplace transform. As indicated in 7.1 this involves complex variable theory (see, for example, [9]), and is beyond the scope of this book. However, in Table 1 of the next chapter we give a short list of inverses which can only be obtained in this way, and which are of importance in the solution of some relatively simple boundary value problems.

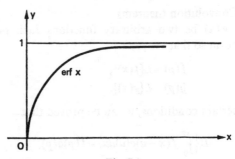

Fig. 7.1

7.4 The error function

One of the functions which will arise in the solution of the heat conduction equation by the Laplace transform method (see Chapter 8) is the error function, erf x, defined by

$$\text{erf } x = \frac{2}{\sqrt{\pi}} \int_0^x e^{-u^2} du. \tag{75}$$

The graph of erf x is shown in Fig. 7.1, the function tending to unity for large x in virtue of the standard result

$$\int_0^\infty e^{-u^2} du = \tfrac{1}{2}\sqrt{\pi}. \tag{76}$$

The complement of the error function, erf x, defined by

$$\text{erfc } x = 1 - \text{erf } x = \frac{2}{\sqrt{\pi}} \int_x^\infty e^{-u^2} du \tag{77}$$

is of particular importance in Chapter 8, and arises in certain boundary value problems in the form erfc $(a/2\sqrt{x})$, where a is a real constant. The Laplace transform of this function can be derived in the following way:

$$L\{\text{erfc}\,(a/2\sqrt{x})\} = \frac{2}{\sqrt{\pi}} \int_0^\infty e^{-px} \left(\int_{a/2\sqrt{x}}^\infty e^{-u^2} du \right) dx \tag{78}$$

$$= \frac{2}{\sqrt{\pi}} \int_0^\infty e^{-u^2} \left(\int_{a^2/4u^2}^\infty e^{-px} dx \right) du \tag{79}$$

$$= \frac{2}{p\sqrt{\pi}} \int_0^\infty e^{-u^2 - a^2 p/4u^2} du. \tag{80}$$

Now writing

$$u^2 + \frac{a^2 p}{4u^2} = \left(u - \frac{a\sqrt{p}}{2u}\right)^2 + a\sqrt{p} \tag{81}$$

and substituting

$$v = u - \frac{a\sqrt{p}}{2u}, \quad \text{or} \quad u = \frac{v + \sqrt{(v^2 + 2a\sqrt{p})}}{2} \tag{82}$$

(80) becomes $\quad \left[\text{since } du = \tfrac{1}{2}dv + \dfrac{v\,dv}{\sqrt{(v^2 + 2a\sqrt{p})}}\right]$

$$L\{\text{erfc}\,(a/2\sqrt{x})\} = \frac{e^{-a\sqrt{p}}}{p\sqrt{\pi}} \int_{-\infty}^{\infty} e^{-v^2}dv + \int_{-\infty}^{\infty} \frac{v}{\sqrt{(v^2 + 2a\sqrt{p})}} e^{-v^2}dv \tag{83}$$

(the range of integration $0 < u < \infty$ implying $-\infty < v < \infty$). The first integral in (83) has the value $\sqrt{\pi}$ in virtue of being twice the integral (76), whilst the second integral is zero since the integrand is an odd function of v.

Hence finally

$$L\{\text{erfc}\,(a/2\sqrt{x})\} = \frac{e^{-a\sqrt{p}}}{p}. \tag{84}$$

7.5 The Heaviside unit step function

The mathematical description of functions possessing one or more finite discontinuities is often simplified by the introduction of the Heaviside unit step function, $H(x)$, defined by

$$H(x) = \begin{cases} 1, & \text{for } x > 0, \\ 0, & \text{for } x < 0. \end{cases} \tag{85}$$

or, more generally

$$H(x - a) = \begin{cases} 1, & \text{for } x > a, \\ 0, & \text{for } x < a. \end{cases} \tag{86}$$

where a is a real constant. The graph of $H(x - a)$ is shown in Fig. 7.2.

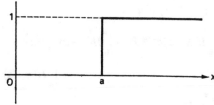

Fig. 7.2

99

As an example of the use of this function consider the square wave pattern shown in Fig. 7.3. If the period of the function is $2a$, and the upper and lower bounds are respectively $y = K$, and $y = -K$ ($K =$ constant), then we have

$$y = \begin{cases} K & \text{for } na < x < (n+1)a, \\ -K & \text{for } (n+1)a < x < (n+2)a, \end{cases} \quad (87)$$

where $n = 0, 2, 4, \ldots$

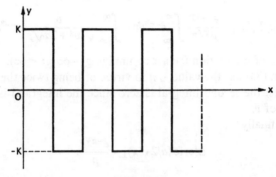

Fig. 7.3

Using the Heaviside function (87) may be written more compactly as one equation:

$$y = K[H(x) - 2H(x-a) + 2H(x-2a) - \ldots]. \quad (88)$$

One further property is of importance. If $y(x)$ is any function then $y(x)H(x)$ is the function $y(x)$ for $x > 0$, and is equal to zero for $x < 0$. Clearly the graph of the function $y(x-a)H(x-a)$ is the graph of the function $y(x)H(x)$ displaced a units to the right (assuming a to be positive).

Using (86) we may now obtain the Laplace transform of the Heaviside function. For, if $a > 0$,

$$L\{H(x-a)\} = \int_0^\infty H(x-a)e^{-px}dx = \int_0^a H(x-a)e^{-px}dx +$$

$$+ \int_a^\infty H(x-a)e^{-px}dx \quad (89)$$

$$= \int_a^\infty e^{-px}dx, \quad (90)$$

100

(since, in $0<x<a$, $H(x-a)=0$, and, in $a<x<\infty$, $H(x-a)=1$). Consequently integrating (90) we have

$$L\{H(x-a)\}=\frac{e^{-pa}}{p}. \tag{91}$$

Similarly

$$L\{H(x-a)y(x-a)\}=\int_0^\infty e^{-px}H(x-a)y(x-a)dx \tag{92}$$

$$=\int_0^a e^{-px}H(x-a)y(x-a)dx+$$
$$+\int_a^\infty e^{-px}H(x-a)y(x-a)dx. \tag{93}$$

The first integral on the right-hand side of (93) is zero. Writing $\tau=x-a$ in the second integral in (93) we finally find

$$L\{H(x-a)y(x-a)\}=e^{-pa}\int_0^\infty e^{-p\tau}y(\tau)d\tau=e^{-pa}\bar{y}(p), \tag{94}$$

where $\bar{y}(p)$ is the Laplace transform of $y(x)$.

7.6 Laplace transforms of derivatives

As a preliminary to the solution of differential equations by the Laplace transform method, we now evaluate the transforms of derivatives of $y(x)$.

Firstly

$$L\left\{\frac{dy}{dx}\right\}=\int_0^\infty e^{-px}\frac{dy}{dx}\,dx=\left(ye^{-px}\right)_0^\infty+p\int_0^\infty ye^{-px}dx \tag{95}$$

$$=-y(0)+pL\{y(x)\} \tag{96}$$
$$=-y(0)+p\bar{y}(p), \tag{97}$$

where $y(0)$ is the value of $y(x)$ at $x=0$, and where in passing from (95) to (96) we have assumed that $ye^{-px}\to0$ as $x\to\infty$.

Similarly

$$L\left\{\frac{d^2y}{dx^2}\right\}=\int_0^\infty e^{-px}\frac{d^2y}{dx^2}\,dx=\left(\frac{dy}{dx}e^{-px}\right)_0^\infty+p\int_0^\infty\frac{dy}{dx}e^{-px}dx. \tag{98}$$

101

Using (97) this becomes $\left(\text{assuming } \dfrac{dy}{dx} e^{-px} \to 0 \text{ as } x \to \infty\right)$

$$L\left\{\frac{d^2y}{dx^2}\right\} = p^2\bar{y}(p) - py(0) - y^{(1)}(0), \tag{99}$$

where $y^{(1)}(0)$ is the value of $\dfrac{dy}{dx}$ at $x = 0$.

Higher derivatives may be transformed in a similar way and, in general, we find

$$L\left\{\frac{d^ny}{dx^n}\right\} = p^n\bar{y}(p) - p^{n-1}y(0) - p^{n-2}y^{(1)}(0) - \ldots - y^{(n-1)}(0), \tag{100}$$

where $y^{(r)}(0)$ is the value of $\dfrac{d^ry}{dx^r}$ at $x = 0$.

Using Theorem 2 of 7.2 we may also obtain the transforms of functions of the type

$$x^m \frac{d^ny}{dx^n} \quad (m, n \text{ positive integers}). \tag{101}$$

For example, using (46) and (97),

$$L\left\{x \frac{dy}{dx}\right\} = -\frac{d}{dp}[p\bar{y}(p) - y(0)] \tag{102}$$

$$= -p\frac{d\bar{y}(p)}{dp} - \bar{y}(p). \tag{103}$$

Likewise

$$L\left\{x \frac{d^2y}{dx^2}\right\} = -\frac{d}{dp}[p^2\bar{y}(p) - py(0) - y^{(1)}(0)] \tag{104}$$

$$= -p^2\frac{d\bar{y}(p)}{dp} - 2p\bar{y}(p) + y(0). \tag{105}$$

This last result will be required in Example 8 of the next section.

7.7 Solution of ordinary differential equations

The essential process involved in solving ordinary linear differential equations by the Laplace transform method is to first convert the equation in $y(x)$ into an equation in $\bar{y}(p)$ using the results of the last section, and then, if possible, to solve for $\bar{y}(p)$. The inversion of

$\bar{y}(p)$ will then give the solution $y(x)$ of the original equation. This technique clearly has advantages only if the equation for $\bar{y}(p)$ is easier to solve than the equation for $y(x)$. In the case of an equation with constant coefficients, the transformed equation for $\bar{y}(p)$ turns out to be an algebraic one, and the Laplace transform method is therefore a powerful tool for the solution of this type of equation as shown by the following two examples.

Example 6. To solve

$$\frac{dy}{dx} + 2y = \cos x \qquad (106)$$

given $y(0) = 1$. Taking the Laplace transform of (106) using (97) and (28) we get

$$p\bar{y}(p) - y(0) + 2\bar{y}(p) = \frac{p}{p^2 + 1}, \qquad (107)$$

which is an algebraic equation for $\bar{y}(p)$. Rearranging (107) gives

$$\bar{y}(p) = \frac{p}{(p+2)(p^2+1)} + \frac{y(0)}{p+2} \qquad (108)$$

$$= \frac{p}{(p+2)(p^2+1)} + \frac{1}{p+2}, \quad \text{(since } y(0) = 1\text{).} \qquad (109)$$

We now invert this function by first writing

$$\frac{p}{(p+2)(p^2+1)} = \frac{A}{p+2} + \frac{Bp+C}{p^2+1}, \qquad (110)$$

where A, B and C are constants. Proceeding in the usual way we find $A = -B = -\frac{2}{5}$, and $C = \frac{1}{5}$, which lead to

$$\bar{y}(p) = -\frac{2}{5}\left(\frac{1}{p+2}\right) + \frac{1}{5}\left(\frac{1+2p}{p^2+1}\right) + \frac{1}{p+2} \qquad (111)$$

$$= \frac{1}{5}\left(\frac{1}{p^2+1}\right) + \frac{2}{5}\left(\frac{p}{p^2+1}\right) + \frac{3}{5}\left(\frac{1}{p+2}\right). \qquad (112)$$

Consequently

$$y(x) = L^{-1}\{\bar{y}(p)\} = \frac{1}{5}\sin x + \frac{2}{5}\cos x + \frac{3}{5}e^{-2x}. \qquad (113)$$

Example 7. To solve the equation

$$\frac{d^2y}{dx^2} + \omega^2 y = kH(x-a), \qquad (114)$$

103

where $\omega(\neq 0)$, k and a are constants and $H(x-a)$ is the Heaviside unit step function.

Taking the Laplace transform of (114) using (99) and (91) we have

$$p^2\bar{y}(p) - py(0) - y^{(1)}(0) + \omega^2\bar{y}(p) = k\frac{e^{-pa}}{p}. \tag{115}$$

Hence

$$\bar{y}(p) = y(0)\frac{p}{p^2+\omega^2} + y^{(1)}(0)\frac{1}{p^2+\omega^2} + ke^{-pa}\frac{1}{p(p^2+\omega^2)} \tag{116}$$

$$= y(0)\frac{p}{p^2+\omega^2} + y^{(1)}(0)\frac{1}{p^2+\omega^2} + \frac{k}{\omega^2}\left(\frac{e^{-pa}}{p} - \frac{e^{-pa}p}{p^2+\omega^2}\right). \tag{117}$$

Consequently we find

$$y(x) = y(0)\cos\omega x + \frac{y^{(1)}(0)}{\omega}\sin\omega x + \frac{k}{\omega^2}H(x-a) -$$

$$- \frac{k}{\omega^2}H(x-a)\cos\omega(x-a), \tag{118}$$

where (94) has been used to invert the last term on the right in (117).

Two general points are worth noting here. Firstly, the solution of an ordinary differential equation obtained by the Laplace transform method is not constructed artificially from the sum of a complementary function and a particular integral (as in the D-operator method), but arises naturally with the boundary conditions already imposed.

Secondly, because of the appearance of the terms $y(0)$, $y^{(1)}(0)$, $\ldots$, $y^{(r)}(0)$ in the transforms of the derivatives of $y(x)$, it is clear that the Laplace transform method is most suited to initial value problems (that is, where the boundary conditions are imposed at $x = 0$).

In the last two examples we have demonstrated the use of the Laplace transform method in the solution of linear equations with constant coefficients. The next example illustrates the use of the method in solving an equation with variable coefficients, although, in general, only equations whose coefficients are polynomials in the independent variable x can be treated in this way.

Example 8. To solve the zero order Bessel equation

$$x\frac{d^2y}{dx^2} + \frac{dy}{dx} + xy = 0 \quad \text{(see Chapter 5).} \tag{119}$$

104

Taking the Laplace transform of (119) using (105) and (46) we find

$$\left[-p^2 \cdot \frac{d\bar{y}(p)}{dp} - 2p\bar{y}(p) + y(0) \right] + [p\bar{y}(p) - y(0)] - \frac{d\bar{y}(p)}{dp} = 0, \quad (120)$$

which may be written as

$$(p^2 + 1) \frac{d\bar{y}(p)}{dp} + p\bar{y}(p) = 0. \quad (121)$$

Hence, integrating, we find

$$\log_e \bar{y}(p) + \tfrac{1}{2} \log_e (p^2 + 1) = \log_e A, \quad (122)$$

where A is an arbitrary integration constant. Accordingly

$$\bar{y}(p) = \frac{A}{\sqrt{(1 + p^2)}}. \quad (123)$$

Inverting with the help of (37) we have finally

$$y(x) = A J_0(x), \quad (124)$$

where $J_0(x)$ is the zero order Bessel function.

It is worth remarking here that if we transform an nth order differential equation whose coefficients are polynomials of degree m, then in virtue of Theorem 2 the transformed equation will be a differential equation of order m. Clearly no advantage is gained if the transformed equation is of higher order than the original equation. However, if $m < n$ the Laplace transform method can usually be used with profit. Even when $m = n$, the transformed equation may be easier to solve than the original equation.

PROBLEMS 7

1. Verify the following results:

(a) $L\{a + bx\} = \dfrac{ap + b}{p^2}$, (b) $L\{x^{-1/2}\} = \sqrt{\left(\dfrac{\pi}{p}\right)}$,

(c) $L\{x \cos ax\} = \dfrac{p^2 - a^2}{(p^2 + a^2)^2}$,

(d) $L\{e^{ax} x^{-1/2}(1 + 2ax)\} = \dfrac{p\sqrt{\pi}}{(p - a)^{3/2}}$,

where a and b are constants.

2. Find the inverse Laplace transform of each of the following functions:

   (a) $\dfrac{p^2}{(p^2+1)\,(p^2+2)}$, (b) $\dfrac{1}{p(p-1)^3}$, (c) $\dfrac{e^{-p}}{p+1}$, (d) $\dfrac{1-e^{-p}}{p}$.

3. Show that the Laplace transform of e^{-x^2} is

$$\tfrac{1}{2}\sqrt{(\pi)}\,e^{p^2/4}\left(1-\operatorname{erf}\frac{p}{2}\right).$$

4. Show that $L\{x^\nu\} = \dfrac{\Gamma(\nu+1)}{p^{\nu+1}}$, where ν is a real constant > -1, and $\Gamma(\nu+1)$ is the gamma function defined by

$$\Gamma(\nu) = \int_0^\infty x^{\nu-1} e^{-x} dx \text{ for } \nu > 0.$$

5. Show, by using the convolution theorem, that if $u(x)$ satisfies the integral equation

$$u(x) = f(x) + \int_0^x g(x-\xi) u(\xi)\, d\xi$$

then

$$\bar{u}(p) = \frac{\bar{f}(p)}{1-\bar{g}(p)},$$ where $\bar{f}(p)$, $\bar{g}(p)$ are respectively the Laplace transforms of $f(x)$ and $g(x)$. Hence obtain the solution of the equation

$$u(x) = \sin 2x + \int_0^x \sin (x-\xi) u(\xi)\, d\xi.$$

6. Solve the following equations by the Laplace transform method:

   (a) $\dfrac{d^2y}{dx^2} + 2\dfrac{dy}{dx} + 2y = 0$, given $y(0)=1$, $y'(0)=-1$.

   (b) $\dfrac{dy}{dx} + y = f(x)$, given $y(0)=1$.

7. Show, using the convolution theorem, that the solution of the equation

$$\frac{d^2y}{dx^2} + 4y = f(x),$$

where $f(x)$ is an arbitrary function and where $y(0)=1$, $y'(0)=1$, is

$$y(x) = \cos 2x + \tfrac{1}{2}\sin 2x + \tfrac{1}{2}\int_0^x f(\xi)\sin 2(x-\xi)\, d\xi.$$

106

CHAPTER 8

Transform Methods for Boundary Value Problems

8.1 Introduction

We now apply the methods of the previous chapter to the solution of boundary value problems.

The basic aim of the transform method is to reduce a given partial differential equation and its boundary conditions to a simpler system by applying an integral transform with respect to one of the independent variables. Since the range of integration in each of the transforms mentioned in Chapter 7 (i.e. the Laplace and Fourier transforms) is infinite, only those independent variables for which the dependent variable of the partial differential equation is defined over an *infinite* range are suitable as variables of transformation.

In this chapter particular use will be made of the Laplace transform since, as we have already seen, it is especially suited to problems where boundary conditions are given at $t=0$ (i.e. Cauchy-type conditions). Such problems arise naturally in the solution of the heat conduction and wave equations, the independent variable t being interpreted there as the time variable. The Laplace transform, however, is not suited to the solution of Laplace's equation. To understand why this is we recall that in order to obtain an unique solution of Laplace's equation within some region R, boundary values must be given at *every* point of the bounding curve of R (the Dirichlet problem – see Chapter 2, 2.4). If, for example, R is the region $x \geq 0$, $0 \leq y \leq a$, where a is some constant then we may attempt to solve $\nabla^2 u = 0$ in R by applying a Laplace transform with respect to the x-variable (since its range is infinite). Boundary conditions on $y=0$, $y=a$ are assumed given. The boundary conditions on $x=0$ are taken care of by the Laplace transform formulae. However, for an unique solution to exist, boundary conditions on the remaining part of the boundary of R – that is, at $x = \infty$ (loosely speaking) – must be given, and these cannot be incorporated into the Laplace transform

107

formulae. For the solution of such problems, the Fourier sine and cosine transforms of Chapter 7, 7.1 are often more appropriate and a few examples of the use of these transforms are given in the last section of this chapter.

Besides the transforms already mentioned, Hankel and Mellin transforms can also be used with profit, the Hankel transform being especially useful in the solution of boundary value problems involving cylindrical symmetry. For a discussion of these transforms the reader should consult [8].

As we shall see in the next section, applying the Laplace transform to a linear *constant* coefficient partial differential equation in two independent variables leads to an ordinary differential equation. If the solution of this ordinary differential equation can be found and inverted, we then obtain a solution of the original partial differential equation. However, this advantage is not always obtainable, and for partial differential equations with coefficients which are functions of the independent variable selected for transformation, the transformed equation (when it can be obtained) is usually another *partial* differential equation of order equal to or higher than the original equation. No advantage is usually gained, therefore, by applying the Laplace transform to equations of this type (compare with similar remarks made in Chapter 7, 7.7 regarding the solution of ordinary differential equations by the Laplace transform method).

In all that follows, we restrict ourselves for simplicity to partial differential equations (necessarily linear) in two independent variables, although repeated use of the transform method can be made to solve certain types of equations in three or more independent variables (see, for example, Problem 3 at the end of this chapter).

8.2 Applications of the Laplace transform

Suppose $u(x, t)$ is an arbitrary function defined for $a \leqslant x \leqslant b$, $t > 0$, where a, b are arbitrary constants. Then

$$L\left\{\frac{\partial u}{\partial t}\right\} = \int_0^\infty e^{-pt} \frac{\partial u}{\partial t}\, dt = \left(ue^{-pt}\right)_0^\infty + p\int_0^\infty ue^{-pt}dt \tag{1}$$

$$= p\bar{u}(x, p) - u(x, 0), \tag{2}$$

where

$$\bar{u}(x, p) = L\{u(x, t)\} \tag{3}$$

is the Laplace transform of $u(x, t)$ with respect to the variable t (assuming that $u(x, t)$ has the necessary properties for the transform to exist).

Similarly we may show that

$$L\left\{\frac{\partial^2 u}{\partial t^2}\right\} = p^2 \bar{u}(x, p) - pu(x, 0) - \frac{\partial}{\partial t} u(x, 0), \qquad (4)$$

where $\frac{\partial}{\partial t} u(x, 0)$ means the partial derivative of $u(x, t)$ with respect to t evaluated at $t = 0$. Equations (2) and (4) are the immediate extensions of equations (97) and (99) of Chapter 7 to the case of a function of two variables.

On the other hand

$$L\left\{\frac{\partial u}{\partial x}\right\} = \int_0^\infty e^{-pt} \frac{\partial u}{\partial x} dt = \frac{d}{dx} \int_0^\infty ue^{-pt}dt \qquad (5)$$

$$= \frac{d}{dx} \bar{u}(x, p), \qquad (6)$$

and

$$L\left\{\frac{\partial^2 u}{\partial x^2}\right\} = \int_0^\infty e^{-pt} \frac{\partial^2 u}{\partial x^2} dt = \frac{d^2}{dx^2} \int_0^\infty ue^{-pt}dt \qquad (7)$$

$$= \frac{d^2}{dx^2} \bar{u}(x, p). \qquad (8)$$

With these results to hand we now illustrate by the following examples the application of the Laplace transform to boundary value problems.

Example 1. To solve the equation

$$\frac{\partial u}{\partial t} + x \frac{\partial u}{\partial x} = x \quad (x > 0, \, t > 0), \qquad (9)$$

where $u = u(x, t)$, given the boundary conditions

$$u(x, 0) = 0 \quad \text{for } x > 0, \qquad (10)$$

and

$$u(0, t) = 0 \quad \text{for } t > 0. \qquad (11)$$

Since both x and t have infinite ranges we may take the Laplace transform with respect to either variable.

109

Consider first taking the Laplace transform with respect to t. Then, using (2) and (6), (9) becomes

$$p\bar{u}(x, p) - u(x, 0) + x \frac{d}{dx}\bar{u}(x, p) = \frac{x}{p},\qquad(12)$$

which, with (10), may be written as

$$\frac{d\bar{u}}{dx} + \frac{p}{x}\bar{u} = \frac{1}{p}.\qquad(13)$$

Solving (13) with the help of the integrating factor x^p we have

$$\bar{u}(x, p) = \frac{A}{x^p} + \frac{x}{p(p+1)},\qquad(14)$$

where A is a constant of integration.

Now since, by (11), $u(0, t) = 0$ we have

$$\bar{u}(0, p) = \int_0^\infty u(0, t)e^{-pt}dt = 0.\qquad(15)$$

Accordingly we must put $A = 0$ in (14) giving the solution

$$\bar{u}(x, p) = \frac{x}{p(p+1)} = x\left(\frac{1}{p} - \frac{1}{p+1}\right).\qquad(16)$$

Inverting (16) with respect to p using the standard results of Chapter 7 we find

$$u(x, t) = xL^{-1}\left\{\frac{1}{p}\right\} - xL^{-1}\left\{\frac{1}{p+1}\right\} = x(1 - e^{-t}),\qquad(17)$$

which is the solution of (9) satisfying the boundary conditions (10) and (11).

We now attempt to solve the problem by taking the Laplace transform of (9) with respect to x.

Writing

$$\bar{u}(p, t) = \int_0^\infty u(x, t)e^{-px}\,dx,\qquad(18)$$

(9) becomes

$$\frac{\partial}{\partial t}\bar{u}(p, t) - \frac{\partial}{\partial p}[p\bar{u}(p, t) - u(0, t)] = \frac{1}{p^2},\qquad(19)$$

where we have used the obvious extension of equation (102) of

110

Chapter 7 to obtain $L\left\{x\dfrac{\partial u}{\partial x}\right\}$. Equation (19) is no simpler than the original equation, and consequently no advantage is gained by taking the Laplace transform of (9) with respect to x. The reason for this lack of simplification is due to the fact that the coefficient of $\dfrac{\partial u}{\partial x}$ in (9) is a function of the variable with respect to which the transform is carried out (see the remarks of (8.1)).

Example 2. To solve the equation

$$x\frac{\partial u}{\partial t}+\frac{\partial u}{\partial x}=x \quad (x>0,\ t>0), \tag{20}$$

given

$$u(x, 0)=0 \quad \text{for } x>0, \tag{21}$$

and

$$u(0, t)=0 \quad \text{for } t>0. \tag{22}$$

Taking the Laplace transform of (20) with respect to t

$$x[p\bar{u}(x, p)-u(x, 0)]+\frac{d\bar{u}(x, p)}{dx}=\frac{x}{p} \tag{23}$$

which, with (21), becomes

$$\frac{d\bar{u}}{dx}+xp\bar{u}=\frac{x}{p}. \tag{24}$$

The solution of (24) is easily found to be

$$\bar{u}(x, p)=\frac{1}{p^2}+Ae^{-(1/2)x^2p}, \tag{25}$$

where A is an arbitrary constant of integration. Now since, by (22), $u(0, t)=0$ we also have

$$\bar{u}(0, p)=\int_0^\infty u(0, t)e^{-pt}dt=0. \tag{26}$$

Putting $x=0$ in (25) and using (26) we find

$$A=-\frac{1}{p^2}, \tag{27}$$

and therefore

$$\bar{u}(x, p)=\frac{1}{p^2}\left(1-e^{-(1/2)x^2p}\right). \tag{28}$$

111

Inverting (28) with respect to p and using the result (Chapter 7, equation (94)) that

$$y(t-a)H(t-a) = L^{-1}\{\bar{y}(p)e^{-ap}\}, \tag{29}$$

where $H(t-a)$ is the Heaviside unit step function, and $\bar{y}(p)$ is the Laplace transform of $y(t)$, we have

$$u(x, t) = L^{-1}\left\{\frac{1}{p^2}\right\} - L^{-1}\left\{\frac{1}{p^2}e^{-(1/2)x^2p}\right\} \tag{30}$$

$$= t - (t - \tfrac{1}{2}x^2)H(t - \tfrac{1}{2}x^2). \tag{31}$$

Accordingly the solution of (20) subject to (21) and (22) is

$$u(x, t) = \begin{cases} t & \text{for } t < \tfrac{1}{2}x^2, \\ \tfrac{1}{2}x^2 & \text{for } t > \tfrac{1}{2}x^2. \end{cases} \tag{32}$$

Example 3. To determine a *bounded* solution of the heat conduction equation

$$\frac{\partial^2 u}{\partial x^2} = \frac{1}{k}\frac{\partial u}{\partial t} \quad (x > 0, \ t > 0), \tag{33}$$

given $u(x, 0) = 0$, and $u(0, t) = u_0$, where u_0 is a constant.

Taking the Laplace transform of (33) with respect to t we have

$$\frac{d^2}{dx^2}\bar{u}(x, p) = \frac{1}{k}[p\bar{u}(x, p) - u(x, 0)] \tag{34}$$

$$= \frac{p}{k}\bar{u}(x, p), \quad \text{(using } u(x, 0) = 0\text{).} \tag{35}$$

which gives

$$\bar{u}(x, p) = Ae^{x\sqrt{(p/k)}} + Be^{-x\sqrt{(p/k)}}, \tag{36}$$

where A and B are constants of integration. Since we require a bounded solution $u(x, t)$, we must also require $\bar{u}(x, p)$ to be bounded. Accordingly the constant A in (36) must be chosen to be zero. Furthermore, since $u(0, t) = u_0$, we have

$$\bar{u}(0, p) = \int_0^\infty u_0 e^{-pt}dt = \frac{u_0}{p}. \tag{37}$$

The solution (36) therefore finally takes the form

$$\bar{u}(x, p) = \frac{u_0}{p}e^{-x\sqrt{(p/k)}}. \tag{38}$$

112

Using the result of Chapter 7, equation (84) (with a slight change of notation) we may invert (38) with respect to p to get

$$u(x, t) = u_0 L^{-1} \left\{ \frac{e^{-(x/\sqrt{k})\sqrt{p}}}{p} \right\} \tag{39}$$

$$= u_0 \operatorname{erfc}\left(\frac{x}{2\sqrt{(kt)}}\right) \tag{40}$$

$$= u_0 \left(1 - \frac{2}{\sqrt{\pi}} \int_0^{x/2\sqrt{kt}} e^{-\lambda^2} d\lambda \right). \tag{41}$$

In each of the preceding examples, the inversion of $\bar{u}(x, p)$ was a relatively easy matter. However, in general, even comparatively simple boundary value problems give rise to functions which are too difficult to invert using the elementary methods described in Chapter 7. For such functions inversions must be carried out by contour integration, as mentioned earlier. A few inverse transforms which are best obtained in this way, and which are of importance in the following examples, are given in Table 1, the inverse being with respect to p. These results may be verified by taking the Laplace transforms of the $u(x, t)$.

Table 1

$\bar{u}(x, p)$	$u(x, t)$
$\dfrac{\cosh x\sqrt{p}}{p \cosh a\sqrt{p}}$	$1 + \dfrac{4}{\pi} \sum_{n=1}^{\infty} \dfrac{(-1)^n}{2n-1} e^{-(2n-1)^2\pi^2 t/4a^2} \cos \dfrac{(2n-1)\pi x}{2a}$
$\dfrac{\sinh px}{p^2 \cosh pa}$	$x + \dfrac{8a}{\pi^2} \sum_{n=1}^{\infty} \dfrac{(-1)^n}{(2n-1)^2} \sin \dfrac{(2n-1)\pi x}{2a} \cos \dfrac{(2n-1)\pi t}{2a}$
$\dfrac{J_0(ix\sqrt{p})}{pJ_0(ia\sqrt{p})}$	$1 - 2 \sum_{n=1}^{\infty} \dfrac{e^{-\lambda_n^2 t/a^2} J_0(\lambda_n x/a)}{\lambda_n J_1(\lambda_n)},$

$\lambda_1, \lambda_2, \ldots$ being the positive roots of $J_0(\lambda) = 0$.

Example 4. To solve the equation

$$\frac{\partial^2 u}{\partial x^2} = \frac{1}{k} \frac{\partial u}{\partial t} \tag{42}$$

113

for $t>0$, $0<x<a$ (where a and k are constants), subject to the boundary conditions

$$u(x, 0) = 0, \quad 0<x<a, \tag{43}$$

$$\left(\frac{\partial u}{\partial x}\right)_{x=a} = 0, \quad t>0, \tag{44}$$

and

$$u(0, t) = u_0, \quad t>0, \tag{45}$$

where u_0 is a constant.

We first remark that in solving the heat conduction equation (42) it is often more convenient to solve the related equation

$$\frac{\partial^2 u}{\partial x^2} = \frac{\partial u}{\partial t} \tag{46}$$

and then to replace t by kt in the resulting solution. Adopting this technique here and taking the Laplace transform of (46) with respect to t, we find

$$\frac{d^2}{dx^2} \bar{u}(x, p) = p\bar{u}(x, p) - u(x, 0) \tag{47}$$

$$= p\bar{u}(x, p), \quad \text{(using (43))}. \tag{48}$$

Hence, solving (48), we have

$$\bar{u}(x, p) = A \cosh x\sqrt{p} + B \sinh x\sqrt{p}, \tag{49}$$

where A and B are constants of integration.

Now since by (45), $u(0, t) = u_0$, we have

$$\bar{u}(0, p) = \int_0^\infty u(0, t)e^{-pt}dt = \frac{u_0}{p}. \tag{50}$$

Likewise, using (44)

$$\left(\frac{\partial \bar{u}}{\partial x}\right)_{x=a} = \int_0^\infty \left(\frac{\partial u}{\partial x}\right)_{x=a} e^{-pt}dt = 0. \tag{51}$$

Using (50) and (51) to determine the constants A and B in (49) we finally find

$$\bar{u}(x, p) = u_0 \left[\frac{\cosh (a-x)\sqrt{p}}{p \cosh a\sqrt{p}}\right]. \tag{52}$$

Inverting with the help of the first result of Table 1, we have

$$u(x, t) = u_0 L^{-1} \left\{ \frac{\cosh (a - x)\sqrt{p}}{p \cosh a\sqrt{p}} \right\} \tag{53}$$

$$= u_0 \left[1 + \frac{4}{\pi} \sum_{n=1}^{\infty} \frac{(-1)^n}{2n - 1} e^{-(2n-1)^2 \pi^2 t / 4a^2} \right.$$

$$\left. \cos \frac{(2n - 1)(a - x)\pi}{2a} \right]. \tag{54}$$

Replacing t by kt and expanding the cosine term, (54) becomes

$$u(x, t) = u_0 \left[1 - \frac{4}{\pi} \sum_{n=1}^{\infty} \frac{1}{2n - 1} e^{-(2n-1)^2 \pi^2 kt / 4a^2} \sin \left(\frac{2n - 1}{2a} \right) \pi x \right], \tag{55}$$

which is the solution of (42) subject to the boundary conditions (43)–(45).

This result can also be obtained by Fourier's method (see Chapter 4).

Example 5. To solve the equation

$$\frac{\partial^2 u}{\partial x^2} = \frac{1}{c^2} \frac{\partial^2 u}{\partial t^2} \tag{56}$$

for $t > 0$, $0 < x < a$ (where a and c are constants), given

$$u(x, 0) = 0, \qquad 0 < x < a, \tag{57}$$

$$\left(\frac{\partial u}{\partial t} \right)_{t=0} = 0, \qquad 0 < x < a, \tag{58}$$

$$u(0, t) = 0, \qquad t > 0, \tag{59}$$

$$\left(\frac{\partial u}{\partial x} \right)_{x=a} = F, \qquad t > 0, \tag{60}$$

where F is a constant.

Writing t for ct in (56) and taking the Laplace transform of the resulting equation

$$\frac{\partial^2 u}{\partial x^2} = \frac{\partial^2 u}{\partial t^2} \tag{61}$$

115

with respect to t, we find (using (4))

$$\frac{d^2}{dx^2}\bar{u}(x, p) = p^2\bar{u}(x, p) - pu(x, 0) - \left(\frac{\partial u}{\partial t}\right)_{t=0} \tag{62}$$

$$= p^2\bar{u}(x, p), \quad \text{(using (57) and (58))}. \tag{63}$$

Solving (63) we get

$$\bar{u}(x, p) = Ae^{px} + Be^{-px}, \tag{64}$$

where A and B are constants of integration.

From (59) and (60) we obtain, respectively,

$$\bar{u}(0, p) = \int_0^\infty u(0, t)e^{-pt}\,dt = 0, \tag{65}$$

and

$$\left(\frac{\partial \bar{u}}{\partial x}\right)_{x=a} = \int_0^\infty \left(\frac{\partial u}{\partial x}\right)_{x=a} e^{-pt}dt = \frac{F}{p}. \tag{66}$$

Inserting these last two results into (64) we find

$$A + B = 0, \tag{67}$$

and

$$(Ae^{pa} - Be^{-pa}) = \frac{F}{p^2}, \tag{68}$$

whence

$$A = -B = \frac{F}{2p^2\cosh pa}. \tag{69}$$

Hence

$$\bar{u}(x, p) = F\left(\frac{\sinh px}{p^2\cosh pa}\right), \tag{70}$$

and accordingly

$$u(x, t) = FL^{-1}\left\{\frac{\sinh px}{p^2\cosh pa}\right\} \tag{71}$$

$$= F\left[x + \frac{8a}{\pi^2}\sum_{n=1}^\infty \frac{(-1)^n}{(2n-1)^2}\sin\frac{(2n-1)\pi x}{2a}\cos\frac{(2n-1)\pi t}{2a}\right] \tag{72}$$

(using the second result of Table 1). Replacing t by ct in (72), we finally obtain the solution of (56) subject to (57)–(60) as

$$u(x, t) = Fa\left[\frac{x}{a} + \frac{8}{\pi^2}\sum_{n=1}^\infty \frac{(-1)^n}{(2n-1)^2}\sin\frac{(2n-1)\pi x}{2a}\cos\frac{(2n-1)\pi ct}{2a}\right]. \tag{73}$$

116

Example 6. To find a *bounded* solution of

$$\frac{\partial^2 u}{\partial r^2} + \frac{1}{r} \frac{\partial u}{\partial r} = \frac{1}{k} \frac{\partial u}{\partial t}, \tag{74}$$

for $t > 0$, $0 \leqslant r < a$ (where a and k are constants), given

$$u(r, 0) = 0, \quad 0 < r < a, \tag{75}$$

and

$$u(a, t) = u_0, \quad t > 0, \tag{76}$$

where u_0 is a constant.

Again writing kt for t and transforming the modified form of (74) we find

$$\frac{d^2}{dr^2} \bar{u}(r, p) + \frac{1}{r} \frac{d}{dr} \bar{u}(r, p) = p\bar{u}(r, p) - u(r, 0), \tag{77}$$

$$= p\bar{u}(r, p), \quad \text{(using (75))}. \tag{78}$$

Equation (78) may be written as

$$\left(r^2 \frac{d^2}{dr^2} + r \frac{d}{dr} - r^2 p \right) \bar{u}(r, p) = 0 \tag{79}$$

which is a Bessel-type equation with the solution

$$\bar{u}(r, p) = A J_0(ir\sqrt{p}) + B Y_0(ir\sqrt{p}), \tag{80}$$

where A and B are constants of integration, and where J_0 and Y_0 are the Bessel functions of zero order of the first and second kind respectively (see Chapter 5).

Since we require a *bounded* solution, and Y_0 is known to be divergent at $r = 0$, we must put $B = 0$ in (80). Furthermore, using (76),

$$\bar{u}(a, p) = \int_0^\infty u(a, t) e^{-pt} dt = \frac{u_0}{p}. \tag{81}$$

Hence finally

$$A = \frac{u_0}{p} \frac{1}{J_0(ia\sqrt{p})} \tag{82}$$

and

$$\bar{u}(r, p) = u_0 \left[\frac{J_0(ir\sqrt{p})}{p J_0(ia\sqrt{p})} \right]. \tag{83}$$

117

The inversion of (83) leads (by the third result of Table 1) to

$$u(r, t) = u_0 L^{-1} \left\{ \frac{J_0(ir\sqrt{p})}{pJ_0(ia\sqrt{p})} \right\} \tag{84}$$

$$= u_0 \left[1 - 2 \sum_{n=1}^{\infty} \frac{e^{-\lambda_n^2 t/a^2} J_0(\lambda_n r/a)}{\lambda_n J_1(\lambda_n)} \right], \tag{85}$$

where λ_1, λ_2, ..., λ_n, ... are the positive roots of the equation $J_0(\lambda) = 0$. Equation (85), with t replaced by kt in the usual way, is then the solution of (74) subject to the boundary conditions (75) and (76).

8.3 Applications of the Fourier sine and cosine transforms

As we have seen in Chapter 7, 7.1, the inverses of the Fourier sine and cosine transforms are given by real integrals, whilst the inverse of the Laplace transform involves an integration in the complex plane. For this reason the sine and cosine transforms are often more convenient to use, although whether or not they can be applied to a given problem depends entirely on the nature of the partial differential equation and the boundary conditions. We now derive some basic results associated with these transforms.

Suppose $u(x, t)$ is a function defined for $0 < x < \infty$, $t > 0$. Then the Fourier sine transform of $u(x, t)$ with respect to x is defined by

$$\bar{u}_s(p, t) = \int_0^\infty u(x, t) \sin px\, dx \tag{86}$$

with an inverse transform

$$u(x, t) = \frac{2}{\pi} \int_0^\infty \bar{u}_s(p, t) \sin px\, dp \tag{87}$$

(compare with equation (6) of Chapter 7). Similarly we define the Fourier cosine transform of $u(x, t)$ with respect to x to be

$$\bar{u}_c(p, t) = \int_0^\infty u(x, t) \cos px\, dx \tag{88}$$

with an inverse transform

$$u(x, t) = \frac{2}{\pi} \int_0^\infty \bar{u}_c(p, t) \cos px\, dp \tag{89}$$

(compare with equation (4) of Chapter 7).

The sine transform of $\frac{\partial u}{\partial x}$ with respect to x is given by

$$\int_0^\infty \frac{\partial u}{\partial x} \sin px\,dx = [u \sin px]_0^\infty - p \int_0^\infty u(x, t) \cos px\,dx. \tag{90}$$

Provided $u(x, t) \to 0$ as $x \to \infty$ (which is often the case in physical problems) then

$$\int_0^\infty \frac{\partial u}{\partial x} \sin px\,dx = -p\bar{u}_c(p, t). \tag{91}$$

Similarly the cosine transform of $\frac{\partial u}{\partial x}$ with respect to x is

$$\int_0^\infty \frac{\partial u}{\partial x} \cos px\,dx = [u \cos px]_0^\infty + p \int_0^\infty u(x, t) \sin px\,dx \tag{92}$$

$$= p\bar{u}_s(p, t) - u(0, t), \tag{93}$$

again provided $u(x, t) \to 0$ as $x \to \infty$.

The transforms of the second derivative $\frac{\partial^2 u}{\partial x^2}$ may be obtained in a like fashion, and we easily find

$$\int_0^\infty \frac{\partial^2 u}{\partial x^2} \sin px\,dx = -p^2\bar{u}_s(p, t) + pu(0, t), \tag{94}$$

and

$$\int_0^\infty \frac{\partial^2 u}{\partial x^2} \cos px\,dx = -p^2\bar{u}_c(p, t) - \left(\frac{\partial u}{\partial x}\right)_{x=0} \tag{95}$$

where, in addition to assuming that $u(x, t) \to 0$ as $x \to \infty$, we have further assumed that $\frac{\partial u}{\partial x} \to 0$ as $x \to \infty$.

From (94) and (95), we now see that the choice between using a sine transform and a cosine transform for the solution of a second-order partial differential equation depends on the boundary conditions, since the sine transform of $\frac{\partial^2 u}{\partial x^2}$ requires a knowledge of $u(0, t)$, whilst the cosine transform of $\frac{\partial^2 u}{\partial x^2}$ requires $\frac{\partial u}{\partial x}$ to be given at $x = 0$.

The following examples illustrate the use of these two transforms.

119

Example 7. To solve the heat conduction equation

$$\frac{\partial^2 u}{\partial x^2} = \frac{1}{k}\frac{\partial u}{\partial t} \qquad (96)$$

given

$$u(0, t) = u_0 \ (=\text{constant}), \qquad (97)$$

and

$$u(x, 0) = 0. \qquad (98)$$

Since $u(0, t)$ is given by (97), the sine transform is appropriate. Taking the sine transform of (96) using (94) we have

$$-p^2\bar{u}_s(p, t) + pu(0, t) = \frac{1}{k}\frac{d}{dt}\bar{u}_s(p, t) \qquad (99)$$

which with (97) becomes·

$$\frac{d}{dt}\bar{u}_s(p, t) + kp^2\bar{u}_s(p, t) = kpu_0. \qquad (100)$$

The solution of this equation is readily found to be

$$\bar{u}_s(p, t)e^{kp^2t} = kpu_0\int e^{kp^2t}dt + A, \qquad (101)$$

or

$$\bar{u}_s(p, t) = Ae^{-kp^2t} + \frac{u_0}{p}, \qquad (102)$$

where A is a constant of integration. To find A we now use (98), which together with (86) gives

$$\bar{u}_s(p, 0) = \int_0^\infty u(x, 0)\sin px\,dx = 0. \qquad (103)$$

Putting $t = 0$ and using (103) in (102) we find

$$A = -\frac{u_0}{p}. \qquad (104)$$

Hence finally

$$\bar{u}_s(p, t) = \frac{u_0}{p}(1 - e^{-kp^2t}). \qquad (105)$$

The inversion formula for the sine transform (see (87)) now gives the solution

$$u(x, t) = \frac{2u_0}{\pi}\int_0^\infty (1 - e^{-kp^2t})\frac{\sin px}{p}\,dp. \qquad (106)$$

Using the result

$$\int_0^\infty \frac{\sin px}{p}\, dp = \frac{\pi}{2}, \quad (x>0), \tag{107}$$

(106) becomes

$$u(x,\, t) = u_0 \left(1 - \frac{2}{\pi} \int_0^\infty e^{-kp^2 t}\, \frac{\sin px}{p}\, dp \right), \tag{108}$$

which, on expanding $\sin px$ and integrating term-by-term leads to the form

$$u(x,\, t) = u_0 \left(1 - \mathrm{erf}\, \frac{x}{2\sqrt{(kt)}} \right) = u_0\, \mathrm{erfc}\, \frac{x}{2\sqrt{(kt)}} \tag{109}$$

as already found in Example 3.

Example 8. If in Example 7 we had imposed, instead of the boundary condition (97), the condition

$$\left(\frac{\partial u}{\partial x} \right)_{x=0} = -\sigma \quad (\sigma = \text{constant}), \tag{110}$$

then by (95) we see that the cosine transform is applicable. Accordingly taking the cosine transform of (96), using (95) and (110), we have

$$-p^2 \bar{u}_c(p,\, t) + \sigma = \frac{1}{k}\, \frac{d}{dt}\, \bar{u}_c(p,\, t). \tag{111}$$

The solution of (111) is

$$\bar{u}_c(p,\, t) = \frac{\sigma}{p^2} + A e^{-kp^2 t}, \tag{112}$$

where A is a constant of integration. Using the boundary condition (98) we find $\bar{u}_c(p,\, 0) = 0$, and hence that

$$A = -\frac{\sigma}{p^2}. \tag{113}$$

Hence

$$\bar{u}_c(p,\, t) = \frac{\sigma}{p^2}\, (1 - e^{-kp^2 t}), \tag{114}$$

and, by (89),

$$u(x,\, t) = \frac{2\sigma}{\pi} \int_0^\infty (1 - e^{-kp^2 t}) \frac{\cos px}{p^2}\, dp. \tag{115}$$

121

This is the solution of (96) subject to the boundary conditions (110) and (98).

8.4 Inhomogeneous equations

In Examples 1 and 2 of 8.1, we demonstrated the use of the Laplace transform in solving two simple (first-order) inhomogeneous equations. Such equations can arise, as we have seen in Chapter 4, 4.3, when certain types of non-homogeneous boundary conditions are imposed on homogeneous equations. Furthermore, many physical problems lead to second-order inhomogeneous equations. For example, the heat conduction equation with a source term included has the form

$$\frac{1}{k}\frac{\partial u}{\partial t} = \frac{\partial^2 u}{\partial x^2} + f(x, t) \tag{116}$$

whilst the wave equation for a vibrating string acted on by an external force $f(x, t)$ takes the form

$$\frac{1}{c^2}\frac{\partial^2 u}{\partial t^2} = \frac{\partial^2 u}{\partial x^2} + f(x, t), \tag{117}$$

where in both cases $f(x, t)$ is a known function.

Likewise, potential theory, which is fundamental to many aspects of fluid dynamics, gravitational and electromagnetic field theory, is concerned with the solution of Poisson's equation (see Chapter 1, 1.3)

$$\nabla^2 u = f(x, y) \text{ (in 2-dimensions)}, \quad \nabla^2 u = f(x, y, z)$$
$$\text{(in 3-dimensions).} \tag{118}$$

The solution of an inhomogeneous equation subject to boundary conditions is frequently a more difficult task than the solution of a homogeneous equation and can sometimes be obtained by using a device called a *Green's function*. This method is discussed in some detail in the next chapter. However, the Green's function technique is not the *only* technique for the solution of inhomogeneous equations and many may be solved by transform methods similar to those used in the earlier sections of this chapter. We conclude therefore by giving a few examples of this approach.

Example 9. To solve

$$\frac{\partial^2 u}{\partial x \partial t} + \sin t = 0 \tag{119}$$

for $t > 0$, given that

$$u(x, 0) = x, \text{ and } u(0, t) = 0. \tag{120}$$

Taking the Laplace transform of (119) with respect to t we have

$$\frac{d}{dx}[p\bar{u}(x, p) - u(x, 0)] + \frac{1}{1+p^2} = 0 \tag{121}$$

$\left(\text{using (2), and the fact that } L\{\sin t\} = \frac{1}{1+p^2}\right).$

Hence inserting the first boundary condition of (120) into (121) we find

$$\frac{d}{dx}\bar{u}(x, p) = \frac{p}{1+p^2} \tag{122}$$

whence

$$\bar{u}(x, p) = \frac{px}{1+p^2} + A, \tag{123}$$

where A is a constant of integration. To determine A we note that since, by the second boundary condition of (120), $u(0, t) = 0$ we also have $\bar{u}(0, p) = 0$. Accordingly putting $x = 0$ into (123) we have

$$A = 0, \tag{124}$$

and therefore

$$\bar{u}(x, p) = \frac{px}{1+p^2}. \tag{125}$$

The inversion of (125) gives

$$u(x, t) = x \cos t \tag{126}$$

as a solution of (119) satisfying (120).

We note here that (119) could also have been solved by the method of separation of variables. For writing

$$u(x, t) = X(x)T(t) \tag{127}$$

we have

$$\frac{dX}{dx}\frac{dT}{dt} + \sin t = 0 \tag{128}$$

123

which separates to give

$$\frac{dX}{dx} = \lambda, \quad \frac{dT}{dt} = -\frac{1}{\lambda} \sin t, \tag{129}$$

where λ is an arbitrary constant of separation. The solutions of the two equations of (129) are respectively

$$X(x) = \lambda x + A, \quad T(t) = \frac{1}{\lambda} \cos t + B, \tag{130}$$

where A and B are arbitrary constants of integration. From (130) and (127) it follows therefore that

$$u(x, t) = (\lambda x + A) \left(\frac{1}{\lambda} \cos t + B \right) \tag{131}$$

$$= x \cos t + \frac{A}{\lambda} \cos t + B\lambda x + AB. \tag{132}$$

Using the given boundary conditions (120) it is easily found that we must choose $A = B = 0$. Hence again

$$u(x, t) = x \cos t. \tag{133}$$

Example 10. To solve

$$\frac{\partial^2 u}{\partial x^2} = \frac{1}{c^2} \frac{\partial^2 u}{\partial t^2} - k \sin \pi x \tag{134}$$

for $0 < x < 1$, $t > 0$, given

$$u(x, 0) = 0, \quad \left[\frac{\partial}{\partial t} u(x, t) \right]_{t=0} = 0, \tag{135}$$

and

$$u(0, t) = 0, \quad u(1, t) = 0. \tag{136}$$

Taking the Laplace transform of (134) with respect to t we have

$$\frac{d^2}{dx^2} \bar{u}(x, p) = \frac{1}{c^2} \left[p^2 \bar{u}(x,p) - pu(x, 0) - \left(\frac{\partial u}{\partial t} \right)_{t=0} \right] - \frac{k \sin \pi x}{p}, \tag{137}$$

which with the boundary conditions (135) becomes

$$\left(\frac{d^2}{dx^2} - \frac{p^2}{c^2} \right) \bar{u}(x, p) = -\frac{k}{p} \sin \pi x. \tag{138}$$

The solution of (138) is

$$\bar{u}(x, p) = Ae^{(p/c)x} + Be^{-(p/c)x} + \frac{kc}{p(p^2 + \pi^2 c^2)} \sin \pi x, \qquad (139)$$

where A and B are arbitrary constants of integration. Now inserting the two boundary conditions (136) into (139) we easily find that $A = B = 0$, and therefore

$$\bar{u}(x, p) = \frac{kc}{p(p^2 + \pi^2 c^2)} \sin \pi x \qquad (140)$$

$$= \frac{k}{\pi^2 c} \left(\frac{1}{p} - \frac{p}{p^2 + \pi^2 c^2} \right) \sin \pi x. \qquad (141)$$

Inverting (141) we finally obtain

$$u(x, t) = \frac{k}{\pi^2 c} (1 - \cos \pi c t) \sin \pi x \qquad (142)$$

as a solution of (134) subject to (135) and (136).

Example 11. The solution of the two-dimensional Poisson equation

$$\nabla^2 u = f(x, y) \qquad (143)$$

in the region $x > 0$, $0 < y < a$ (where a is a given constant), subject to the boundary conditions

$$u(x, y) \to 0 \text{ as } x \to \infty, \qquad (144)$$

$$\frac{\partial}{\partial x} u(x, y) \to 0 \text{ as } x \to \infty, \qquad (145)$$

and

$$u(0, y) = g(y), \qquad (146)$$

where $g(y)$ is a given function, may sometimes be obtained using a Fourier sine transform. For applying a sine transform with respect to x and using (94) (which is applicable here in virtue of (144) and (145)) we find that (143) becomes

$$-p^2 \bar{u}_s(p, y) + p u(0, y) + \frac{d^2}{dy^2} \bar{u}_s(p, y) = \bar{f}_s(p, y) \qquad (147)$$

where $\bar{f}_s(p, y)$ is the Fourier sine transform of $f(x, y)$ with respect to x defined by

$$\bar{f}_s(p, y) = \int_0^\infty f(x, y) \sin px \, dx. \qquad (148)$$

125

Accordingly, using (146), (147) becomes

$$\left(\frac{d^2}{dy^2} - p^2\right) \bar{u}_s(p, y) = \bar{f}_s(p, y) - pg(y). \tag{149}$$

Now provided $f(x, y)$ is such that $\bar{f}_s(p, y)$ exists (149) may be solved for $\bar{u}_s(p, y)$. This solution will contain two arbitrary constants which will be determined by imposing boundary conditions on the lines $y = 0$ and $y = a$; say

$$u(x, 0) = \lambda(x), \quad u(x, a) = \mu(x), \tag{150}$$

where $\lambda(x)$ and $\mu(x)$ are given functions. The Fourier inversion (see (87)) of $\bar{u}_s(p, y)$ will then give the required solution $u(x, y)$.

PROBLEMS 8

1. Show that the bounded solution of the wave equation

$$\frac{\partial^2 u}{\partial x^2} = \frac{1}{c^2} \frac{\partial^2 u}{\partial t^2}$$

which satisfies the boundary conditions $u(x, 0) = 0$, $\left(\dfrac{\partial u}{\partial t}\right)_{t=0} = 0$ and $u(0, t) = J_0(at)$, where a and c are constants and J_0 is the zero order Bessel function, is

$$u(x, t) = H\left(t - \frac{x}{c}\right) J_0\left[a\left(t - \frac{x}{c}\right)\right],$$

H being the Heaviside unit step function (see Chapter 7, 7.5).

2. Given that

$$L^{-1}\left\{\frac{\sinh x\sqrt{p}}{p \sinh l\sqrt{p}}\right\} = \frac{x}{l} + \frac{2}{\pi} \sum_{n=1}^{\infty} \frac{(-1)^n}{n} e^{-n^2\pi^2 t/l^2} \sin\frac{n\pi x}{l},$$

solve, by the Laplace transform method, the equation

$$\frac{\partial^2 u}{\partial x^2} = \frac{1}{k} \frac{\partial u}{\partial t}$$

subject to the boundary conditions $u(0, t) = u_1$, $u(l, t) = u_2$ and $u(x, 0) = 0$ for $0 < x < l$, where l, u_1 and u_2 are constants.

126

3. Consider the two-dimensional heat conduction equation

$$\frac{\partial^2 u}{\partial x^2} + \frac{\partial^2 u}{\partial y^2} = \frac{1}{k}\frac{\partial u}{\partial t},$$

and the boundary conditions $u(0, y, t) = 0$, $u(\pi, y, t) = 0$, $u(x, 0, t) = C$, $u(x, y, 0) = 0$, where $0 < x < \pi$, $y > 0$ and $t > 0$, C being a constant. Obtain a bounded solution of this boundary value problem by first taking a Laplace transform with respect to t, and then a Fourier sine transform with respect to x. (The result

$$L^{-1}\left\{\frac{e^{-y\sqrt{(p+\alpha)}}}{p}\right\} = \frac{2}{\sqrt{\pi}}\int_{y/2\sqrt{t}}^{\infty} e^{-(u^2 + \alpha y^2/4u^2)}du$$

may be assumed, α being a constant).

CHAPTER 9

Green's Functions and Generalised Functions

9.1 Introduction

As mentioned in Chapter 8, 8.4, a powerful method of solving certain types of linear inhomogeneous equations is based on a device known as a Green's function. We illustrate the first stages of this method by considering the solution of an inhomogeneous ordinary differential equation.

Suppose the equation has the form

$$Lu(x) = f(x), \tag{1}$$

where L is a linear differential operator, and $f(x)$ is a given function (the source term), the solution being required on the interval $0 \leqslant x \leqslant l$ (where l is some constant). Now instead of considering $f(x)$ as a *continuous* source function, we approximate to it by a set of discrete source functions $f(\xi_1), f(\xi_2), \ldots f(\xi_n)$ acting at the points $x = \xi_1, x = \xi_2, \ldots x = \xi_n$, all within $0 \leqslant x \leqslant l$. We define the function $G(x; \xi_k)$ to be the solution of (1) due to a *unit* point source acting at ξ_k. The solution due to the single effect of is therefore $G(x; \xi_k) f(\xi_k)$. Summing such solutions for all the n-point source terms acting on $0 \leqslant x \leqslant l$, the solution takes the form

$$u(x) = \sum_{k=1}^{n} G(x; \xi_k) f(\xi_k). \tag{2}$$

As n becomes larger so the number of point-source functions $f(\xi_n)$ increases and accordingly a better and better approximation to $f(x)$ is cbtained. In the limit as $n \to \infty$, so $\xi_k \to \xi_{k+1}$ (for all k) and the summation in (2) may be replaced by an integral to give the required solution of (1) in the form

$$u(x) = \int_0^l G(x; \xi) f(\xi) d\xi. \tag{3}$$

The function $G(x; \xi)$ is called the Green's function of the problem.

128

Similar results, as we shall see in 9.3, may be obtained for linear partial differential equations; for example, the solution of Poisson's equation in two-dimensions

$$\nabla^2 u = f(x, y) \tag{4}$$

for $0 \leqslant x \leqslant a$, $0 \leqslant y \leqslant b$ may be written as

$$u(x, y) = \int_0^a \int_0^b G(x, y; \xi, \eta) f(\xi, \eta) d\xi d\eta, \tag{5}$$

where $G(x, y; \xi, \eta)$ is the Green's function of the problem.

9.2 Generalised functions

It is clear from 9.1 that the essential problem of developing the solution of inhomogeneous equations is in actually finding the Green's function in each case. This problem is made considerably easier by the use of a new type of function – called a *generalised* function – together with the techniques of the Laplace and Fourier transforms. So that we may understand what is meant by a generalised function consider $\nabla^2 \psi$ in spherical polar coordinates (r, φ, θ) (see Chapter 4, equation (18)). Thence

$$\nabla^2 \psi(r) = \frac{1}{r^2} \frac{d}{dr} \left(r^2 \frac{d\psi(r)}{dr} \right). \tag{6}$$

If $\psi(r) = \frac{1}{r}$, then $\nabla^2 \psi(r) = 0$ for all points where $r \neq 0$. At $r = 0$, however, $\nabla^2 \psi(r)$ does not exist. Now by the divergence theorem (or Gauss' theorem)

$$\int_V \text{div } \mathbf{F} \, dV = \int_S \mathbf{F} \cdot \mathbf{n} \, dS \tag{7}$$

where $\mathbf{F}$ is a differentiable vector field, V the volume of a closed region bounded by a surface S, and $\mathbf{n}$ the outward drawn unit normal at a typical point of S. Consider a sphere of radius a, centre $r = 0$.

Then by (7)

$$\int_V \nabla^2 \left(\frac{1}{r} \right) dV = \int_V \text{div} \left(\text{grad } \frac{1}{r} \right) dV = \int_S \left(\text{grad } \frac{1}{r} \right) \cdot \mathbf{n} \, dS. \tag{8}$$

Now grad $\psi(r, \varphi, \theta)$ in spherical polar coordinates has the form (see, for example, [1], [13])

$$\text{grad } \psi = \mathbf{e}_r \frac{\partial \psi}{\partial r} + \mathbf{e}_\theta \frac{1}{r} \frac{\partial \psi}{\partial \theta} + \mathbf{e}_\phi \frac{1}{r \sin \theta} \frac{\partial \psi}{\partial \varphi}, \qquad (9)$$

where $\mathbf{e}_r$, $\mathbf{e}_\theta$, $\mathbf{e}_\phi$ are unit vectors in the directions of increasing r, θ, φ respectively.

Hence on the surface S of a sphere of radius a (no θ, φ dependence) we have

$$\left[\text{grad } \frac{1}{r} \right]_{r=a} = \mathbf{e}_r \left[\frac{d}{dr} \left(\frac{1}{r} \right) \right]_{r=a} = -\frac{\mathbf{e}_r}{a^2}, \qquad (10)$$

where now $\mathbf{e}_r$ is in the direction of the outward drawn normal $\mathbf{n}$ to the surface of the sphere. Using (8) we finally find that

$$\int_V \nabla^2 \left(\frac{1}{r} \right) dV = \int_S \left(\text{grad } \frac{1}{r} \right) \cdot \mathbf{n} \, dS = -\frac{1}{a^2} \int_S \mathbf{e}_r \cdot \mathbf{n} \, dS = -4\pi, \qquad (11)$$

since $\mathbf{e}_r \cdot \mathbf{n} = 1$.

We see then that $\nabla^2 \left(\frac{1}{r} \right)$ is a function which has the following properties:

  (a) it is undefined at $r = 0$,

and (b) vanishes if $r \neq 0$,

but (c) has an integral over any sphere, centred at $r = 0$, equal to -4π.

Clearly this is no ordinary function. However, the origin of the singular behaviour is that, at $r = 0$, grad $\frac{1}{r}$ is undefined with the consequence that the divergence theorem is not really valid. Nevertheless this does not remove the peculiarity of the behaviour of $\nabla^2 \left(\frac{1}{r} \right)$ since we can approximate more and more closely to $\frac{1}{r}$ by some sequence of functions $\psi_1(r)$, $\psi_2(r)$, ... $\psi_n(r)$ such that all $\psi_n(r)$ are finite and continuous everywhere *including* the origin (see Fig. 9.1). It can be shown that in the limit $n \to \infty$ the divergence theorem still applies giving a volume integral -4π. Hence we are really faced with the situation that

$$\int_V \nabla^2 \left(\frac{1}{r} \right) dV = \begin{cases} -4\pi, & \text{if } V \text{ contains } r = 0, \\ 0, & \text{if } V \text{ does not contain } r = 0. \end{cases} \qquad (12)$$

130

We may express this result more compactly by writing

$$\nabla^2\left(\frac{1}{r}\right) = -4\pi\delta(r), \tag{13}$$

where $\delta(r)$ is a 'generalised' function called the Dirac delta function. For (12) to be valid we must require that the delta function have the property

$$\int_V \delta(r)dV = \begin{cases} 1, & \text{if } V \text{ contains } r=0, \\ 0, & \text{otherwise.} \end{cases} \tag{14}$$

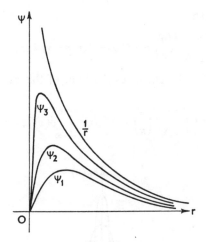

Fig. 9.1

This result is but a special form of the general definition of the delta function in vector form

$$\int_V f(\mathbf{r})\delta(\mathbf{r}-\mathbf{a})dV = \begin{cases} f(\mathbf{a}), & \text{if } V \text{ contains the point } \mathbf{a}, \\ 0, & \text{otherwise,} \end{cases} \tag{15}$$

obtained by putting $f(\mathbf{r})=1$, $\mathbf{a}=0$ in (15).

It is often useful to consider the delta function in its one-dimensional form

$$\int_{-\infty}^{\infty} f(x)\delta(x-a)dx = f(a), \tag{16}$$

where a is a parameter.

For (15) and (16) to be valid results it is certainly *sufficient* that the

131

functions be continuous and infinitely differentiable everywhere. However, less well-behaved functions than these are also allowable (see, for example, Problem 5 at the end of the chapter). Detailed discussions of this topic may be found in [16], [17].

Now although we have seen that no ordinary function has the properties required of the delta function, we may approximate to $\delta(x)$ (in one-dimension) by a sequence of ordinary functions ($n = 1, 2, 3, \ldots$), where

$$\delta_n(x) = \sqrt{\left(\frac{n}{\pi}\right)} e^{-nx^2} \qquad (17)$$

(see Fig. 9.2). Then

$$\int_{-\infty}^{\infty} \sqrt{\left(\frac{n}{\pi}\right)} e^{-nx^2} \, dx = 1, \quad n = 1, 2, 3, \ldots, \qquad (18)$$

and $\delta(x)$ may be thought of as the limit of this sequence as $n \to \infty$ so that

$$\underset{n \to \infty}{Lt} \int_{-\infty}^{\infty} \sqrt{\left(\frac{n}{\pi}\right)} e^{-nx^2} dx = \int_{-\infty}^{\infty} \delta(x) dx = 1 \qquad (19)$$

as required.

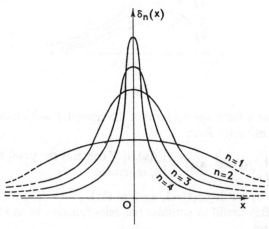

Fig. 9.2

From the diagram we see that the delta function sequence has a physical analogy of an 'impulsive' force acting for a short time only. For example, a unit impulse would be described by a force function

132

$p(t)$, say, which vanishes outside a small time interval Δt centred about $t = 0$ and such that

$$\int_{-\infty}^{\infty} p(t)dt = 1. \tag{20}$$

To idealise this to an instantaneous force we describe it by $\delta(t)$ which vanishes except at one point ($t = 0$) and is such that

$$\int_{-\infty}^{\infty} \delta(t)dt = 1. \tag{21}$$

We now give some examples illustrating certain properties of the delta function.

Example 1. To show that

$$f(x)\delta(x - a) = f(a)\delta(x - a). \tag{22}$$

It must be understood here that properties such as this only have a meaning when each side of the relation is used as an integrand in (16), and thence shown to give the same result.

Now

$$\int_{-\infty}^{\infty} f(a)\delta(x - a)dx = f(a) \int_{-\infty}^{\infty} \delta(x - a)dx = f(a), \tag{23}$$

and $\qquad \int_{-\infty}^{\infty} f(x)\delta(x - a)dx = f(a)$, using (16). $\qquad$ (24)

Hence (23) and (24) are identical, and the result (22) therefore follows.

Example 2. To show that

$$x\delta(x) = 0. \tag{25}$$

Here we have, using (16) with $a = 0$,

$$\int_{-\infty}^{\infty} xf(x)\delta(x)dx = [xf(x)]_{x=0} = 0, \tag{26}$$

whence (25) follows.

Example 3. To show that the delta function is an even function of its argument—that is

$$\delta(a - x) = \delta(x - a). \tag{27}$$

Consider

$$\int_{-\infty}^{\infty} \delta(a - x)f(x)dx = \int_{\infty}^{-\infty} \delta[y - (-a)]f(-y)(-dy), \tag{28}$$

where we have written $y = -x$.

133

Hence

$$\int_{-\infty}^{\infty} \delta(a-x)f(x)dx = \int_{-\infty}^{\infty} \delta[y-(-a)]f(-y)dy$$

$$= f[-(-a)] = f(a) = \int_{-\infty}^{\infty} \delta(x-a)f(x)dx, \qquad (29)$$

using (16) yet again.

Accordingly $\delta(a-x) = \delta(x-a)$ as required to be shown.

Example 4. To show that

$$\delta(x) = \frac{d}{dx} H(x), \qquad (30)$$

where $H(x)$ is the Heaviside unit step function (itself a generalised function) defined by

$$H(x) = \begin{cases} 1, & x>0, \\ 0, & x<0, \end{cases} \qquad (31)$$

(see Chapter 7.7.4).

This follows immediately from the relation

$$\int_{-\infty}^{x} \delta(y)dy = H(x). \qquad (32)$$

Finally we remark that the three-dimensional delta function $\delta(\mathbf{r} - \mathbf{a})$, where $\mathbf{r}$ and $\mathbf{a}$ are vectors, may be written as

$$\delta(\mathbf{r} - \mathbf{a}) = \delta(x-a_1)\delta(y-a_2)\delta(z-a_3), \qquad (33)$$

where $\mathbf{r} = \mathbf{i}x + \mathbf{j}y + \mathbf{k}z$ and $\mathbf{a} = \mathbf{i}a_1 + \mathbf{j}a_2 + \mathbf{k}a_3$.

Besides the delta function and the Heaviside step function, other generalised functions are known (see, for example, Problem 3 at the end of the chapter). The use of various generalised functions in the advanced theory of partial differential equations is discussed at fair length in [16].

9.3 Green's functions

We now show how the Dirac delta function may be used to develop the solution of the inhomogeneous ordinary differential equation

$$Lu(x) = f(x) \qquad (34)$$

discussed in 9.1, where L is a linear differential operator with constant coefficients. For simplicity we assume that (34) is valid for all x, and that no boundary conditions are imposed.

Consider the equation

$$LG(x; x') = \delta(x - x'), \tag{35}$$

where x' is a parameter. When $x \neq x'$, (35) is just the homogeneous form of (34). The function $G(x; x')$ is called the *Green's function* for the operator L, and represents the effect at x due to a delta function acting at x'. Now to solve (34) we multiply (35) through by $f(x')$ and integrate over the whole range $-\infty < x' < \infty$. In this way we easily find

$$\int_{-\infty}^{\infty} LG(x; x')f(x')dx' = \int_{-\infty}^{\infty} f(x')\delta(x - x')dx' = f(x) \tag{36}$$

making use of the defining property of the delta function. Interchanging the order of the differential operator L and the integral sign in (36) (a process which can be justified rigorously) we finally find

$$L\left\{\int_{-\infty}^{\infty} G(x; x')f(x')dx'\right\} = f(x). \tag{37}$$

Comparing (37) with (34) we see that the solution of (34) may be written as

$$u(x) = \int_{-\infty}^{\infty} G(x; x')f(x')dx'. \tag{38}$$

This has the same form as the solution (3) found in 9.1 by a less sophisticated method.

The use of the delta function in this way may be readily extended to certain types of linear partial differential equations in two or more independent variables. Consider the partial differential equation

$$L_x u(\mathbf{x}) = f(\mathbf{x}) \tag{39}$$

where $\mathbf{x}$ is a vector in three-dimensions, and L_x is a linear constant coefficient differential operator in three independent variables. Both u and f are, in general, functions of three variables. As before, consider the equation

$$L_x G(\mathbf{x}; \mathbf{x}') = \delta(\mathbf{x} - \mathbf{x}'). \tag{40}$$

$G(\mathbf{x}; \mathbf{x}')$ is then the Green's function of the problem, and represents the effect at the point $\mathbf{x}$ of a delta function source at the point

$\mathbf{x} = \mathbf{x'}$. Multiplying through by $f(\mathbf{x'})$ and integrating over the volume V of the $\mathbf{x'}$ space, we find

$$L\left\{\int_{V_{\mathbf{x'}}} G(\mathbf{x}; \mathbf{x'})f(\mathbf{x'})dV_{\mathbf{x'}}\right\} = \int_{V_{\mathbf{x'}}} f(\mathbf{x'})\delta(\mathbf{x'} - \mathbf{x})dV_{\mathbf{x'}} = f(\mathbf{x}), \quad (41)$$

using the basic result (15).

Comparing (39) and (41) we see that the solution of (39) may be written as

$$u(\mathbf{x}) = \int_{V_{\mathbf{x'}}} G(\mathbf{x}; \mathbf{x'})f(\mathbf{x'})dV_{\mathbf{x'}}. \quad (42)$$

Clearly (42) is valid no matter how many components $\mathbf{x}$ may have. Accordingly the Green's function technique can be applied, in principle, to any linear constant coefficient inhomogeneous partial differential equation in any number of independent variables. However, although a neat formalism has been developed, in practice the difficulty arises in actually finding the Green's function. As shown by the next example, the Laplace transform is often of use here.

Example 5. Consider the ordinary differential equation

$$\left(\frac{d^2}{dt^2} + \omega_0{}^2\right)u(t) = f(t), \quad (43)$$

where ω_0 is a constant, the solution being required for $0 \leqslant t < \infty$. Then the Green's function $G(t; t')$ for the infinite space solution (that is, all t) is, by definition, the solution of the equation

$$\left(\frac{\partial^2}{\partial t^2} + \omega_0{}^2\right)G(t; t') = \delta(t - t'), \quad (44)$$

where $0 \leqslant t' < \infty$. Now taking the Laplace transform of (44) with respect to t we find (using equation (4) of Chapter 8) that

$$p^2\bar{G}(p; t') - pG(0; t') - G'(0; t') + \omega_0{}^2\bar{G}(p; t') = e^{-pt'}, \quad (45)$$

where

$$G'(0; t') = \left[\frac{\partial G(t; t')}{\partial t}\right]_{t=0}, \quad (46)$$

and the Laplace transform of the delta function is given by

$$\int_0^\infty e^{-pt}\delta(t - t')dt = e^{-pt'}. \quad (47)$$

Accordingly (45) becomes

$$(p^2 + \omega_0{}^2)\bar{G}(p; t') = pG(0; t') + G'(0; t') + e^{-pt'}, \quad (48)$$

whence

$$\bar{G}(p; t') = \frac{p}{p^2 + \omega_0^2} \, G(0; t') + \frac{1}{p^2 + \omega_0^2} \, G'(0; t') + \frac{e^{-pt'}}{p^2 + \omega_0^2}. \quad (49)$$

Inverting, using the result (94) of Chapter 7 on the last term of (49), we find the Green's function

$$G(t; t') = G(0; t') \cos \omega_0 t + \frac{1}{\omega_0} G'(0; t') \sin \omega_0 t +$$

$$+ \frac{1}{\omega_0} \sin \omega_0(t - t') \, . \, H(t - t'), \quad (50)$$

where $H(t - t')$ is the Heaviside unit step function. The solution of (43) is therefore, by comparison with (38),

$$u(t) = \int_0^\infty G(t; t') f(t') dt'$$

$$= \int_0^\infty G(0; t') f(t') \cos \omega_0 t \, dt' + \frac{1}{\omega_0} \int_0^\infty G'(0; t') f(t') \sin \omega_0 t \, dt' +$$

$$+ \frac{1}{\omega_0} \int_0^\infty \sin \omega_0(t - t') H(t - t') f(t') dt' \quad (51)$$

$$= A \cos \omega_0 t + B \sin \omega_0 t + \frac{1}{\omega_0} \int_0^\infty \sin \omega_0(t - t') H(t - t') f(t') dt', \quad (52)$$

where A and B are constants. This is precisely the general solution of (43) made up of the solution $A \cos \omega_0 t + B \sin \omega_0 t$ of the homogeneous form of (43), together with a particular integral of (43) which is the last term of (52).

We have presented here only an introduction to what is really a substantial field of study, but the aim has been to show how the method can be set up with the use of the delta function. The problem of solving for the Green's functions in more complicated cases (for example, for the wave equation in three space dimensions) often involves the use of complex Fourier transforms and the corresponding transforms of certain generalised functions. A discussion of these topics is beyond the scope of the present book.

PROBLEMS 9

1. Show that $\delta(ax) = \dfrac{1}{a}\,\delta(x)$, $a > 0$.

2. Show that $\delta(a^2 - x^2) = \dfrac{1}{2a}\,[\delta(x - a) + \delta(x + a)]$.

3. If
$$|x| = \begin{cases} x, & x > 0 \\ -x, & x < 0 \end{cases} \quad \text{and} \quad \text{sgn}\,(x) = \begin{cases} 1, & x > 0, \\ -1, & x < 0, \end{cases}$$

show that
$$\text{sgn}\,(x) = 2H(x) - 1, \quad \frac{d}{dx}\,\text{sgn}\,(x) = 2\delta(x), \quad \frac{d}{dx}|x| = \text{sgn}(x).$$

4. Assuming that the derivatives of $\delta(x)$ all exist, show that
$$\int_{-\infty}^{\infty} f(x)\delta^{(1)}(x)\,dx = -f^{(1)}(0),$$
and
$$\int_{-\infty}^{\infty} f(x)\delta^{(n)}(x)\,dx = (-1)^n f^{(n)}(0),$$
where $\delta^{(n)}(x)$ is the nth derivative of $\delta(x)$, and $f^{(n)}(0)$ is the nth derivative of $f(x)$ evaluated at $x = 0$.

5. Draw the graph of $f(x) = e^{-|x|}$ and hence verify that the function is not differentiable at $x = 0$. Prove, nevertheless, that
$$\int_{-\infty}^{\infty} e^{-|x|}\delta(x)\,dx = e^0 = 1$$
by approximating to $\delta(x)$ by the sequence $\delta_n(x) = \sqrt{\left(\dfrac{n}{\pi}\right)}e^{-nx^2}$

and considering
$$\underset{n \to \infty}{Lt}\sqrt{\left(\frac{n}{\pi}\right)}\int_{-\infty}^{\infty} e^{-|x|}e^{-nx^2}\,dx.$$

6. Show, by taking the Laplace transform with respect to x of the equation
$$\left(-\frac{d^2}{dx^2} - k^2\right)G(x;\,x') = \delta(x - x'),$$

138

that the Green's function $G(x; x')$ on the interval $(0, 1)$ with the boundary conditions

$$G(0; x') = 0 \quad \text{and} \quad \left[\frac{\partial G(x; x')}{\partial x}\right]_{x=0} = G(1; x')$$

is

$$G(x; x') = \frac{\sin kx \sin k(1 - x')}{k(\sin k - k)} - \frac{\sin k(x - x')}{k} H(x - x').$$

CHAPTER 10

The Numerical Approach

10.1 Introduction

The partial differential equations discussed in the preceding chapters were linear equations, and were capable of exact analytical solution within simple regular-shaped regions such as a circle and a square, or (in three dimensions) within a sphere, cube or cylinder. The boundary conditions in these previous examples were also relatively straightforward and expressible in simple mathematical form. In dealing with many problems arising from the modelling of physical processes, however, the region within which a solution is required may be rather complicated and irregular in shape. Furthermore the boundary conditions may be anything but simple, and in some cases known only as a set of discrete numerical values. To add to these difficulties the equation to be solved may no longer be linear (for example, the Navier−Stokes equations of fluid dynamics, and the Einstein field equations of general relativity which describe the gravitational field, are both examples of *non*-linear partial differential equations). In these circumstances it is usually very difficult, if not impossible, to obtain exact analytical solutions and resort has to be made to finding approximate numerical solutions. Nowadays computer software packages are available for the solution of partial differential equations, but it is clearly important to know and understand the numerical method or algorithm used in a typical package to ensure that it is appropriate to the problem under discussion. Totally spurious results can only too often be obtained because of ignorance of the restrictions of the underlying numerical method. Unfortunately too, there is a real danger that such results, spurious as they may be, will be totally accepted just because they come out of a computer. Computing does not replace mathematics − it only complements it − and a proper understanding of the underlying mathematics, whether it be analytically or numerically based, is essential. Of course, the field of numerical analysis is a large one and in this short chapter it is only possible to give the bare outline of two simple numerical techniques, and to show their application to the solution of two well-known *linear* equations – the Laplace equation and the heat conduction or diffusion equation. For more extensive treatments of numerical methods the reader is referred to the texts [15, 18, 19] listed under Further Reading at the end of this book.

10.2 Finite difference methods

We first recall how to express the derivatives of a function $y(x)$ in terms of its values $y_0, y_1, y_2, \ldots, y_n$ at corresponding x-values $x_0, x_1, \ldots, x_n$. These are assumed to be regularly spaced with a separation h so that

$$x_{r+1} = x_r + h. \tag{1}$$

Now consider a Maclaurin series expressing y_{r+1} in terms of the values of y_r and its derivatives $y_r', y_r'', \ldots$ at x_r. Then

$$y_{r+1} = y_r + h y_r' + \frac{h^2}{2!} y_r'' + \ldots \tag{2}$$

and consequently to terms of order h we have an approximate form for the derivative given by

$$y_r' = \frac{y_{r+1} - y_r}{h}. \tag{3}$$

By writing

$$y_{r-1} = y_r - h y_r' + \frac{h^2}{2!} y_r'' - \ldots \tag{4}$$

and subtracting (4) from (2) we have an alternative form for y_r', namely

$$y_r' = \frac{y_{r+1} - y_{r-1}}{2h}, \tag{5}$$

which is correct to order h^2.

Similarly by adding (2) and (4) we have for the second derivative

$$y_r'' = \frac{y_{r-1} - 2y_r + y_{r+1}}{h^2}, \tag{6}$$

which is also correct to order h^2.

We can now apply these approximations to the solution of Laplace's equation within a square subject to boundary conditions of the Dirichlet type (see Chapter 2, 2.4).

Example 1. Solve the equation

$$\frac{\partial^2 u}{\partial x^2} + \frac{\partial^2 u}{\partial y^2} = 0 \tag{7}$$

within the square region subject to the following boundary conditions:

$$u(x, y) = \begin{array}{l} 0 \text{ when } x = 0, \quad 0 \leqslant y \leqslant 1, \\ 0 \text{ when } x = 1, \quad 0 \leqslant y \leqslant 1, \\ 0 \text{ when } y = 1, \quad 0 \leqslant x \leqslant 1, \\ x(1 - x) \text{ when } y = 0, \quad 0 \leqslant x \leqslant 1. \end{array} \tag{8}$$

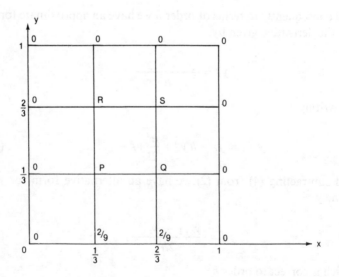

Fig. 10.1.

142

We first divide the square region into nine smaller squares using a rather coarse grid or mesh with a mesh length $h = \frac{1}{3}$ (see Fig. 10.1). At a mesh point (or node) we have $x = mh$, $y = nh$, where m and n are integers. A typical node will be the (m, n) node and the value of $u(x, y)$ at this point will be denoted by $u_{m,n}$. It now follows from the basic result (6) that

$$\left(\frac{\partial^2 u}{\partial x^2}\right)_{m,n} = \frac{u_{m-1,n} - 2u_{m,n} + u_{m+1,n}}{h^2} \tag{9}$$

and

$$\left(\frac{\partial^2 u}{\partial y^2}\right)_{m,n} = \frac{u_{m,n-1} - 2u_{m,n} + u_{m,n+1}}{h^2}. \tag{10}$$

Hence at the (m, n) node the Laplace equation may be approximated by

$$u_{m-1,n} - 2u_{m,n} + u_{m+1,n} + u_{m,n-1} - 2u_{m,n} + u_{m,n+1} = 0 \tag{11}$$

or

$$u_{m-1,n} + u_{m,n+1} + u_{m,n-1} + u_{m,n+1} - 4u_{m,n} = 0 \tag{12}$$

to terms of order h^2.

Consequently

$$u_{m,n} = \frac{1}{4}(u_{m-1,n} + u_{m,n+1} + u_{m,n-1} + u_{m,n+1}) \tag{13}$$

from which it follows that the value of $u(x, y)$ at a typical node is the arithmetic mean of the values of $u(x, y)$ at the four surrounding nodes. Applying (13) at the four interior nodes P, Q, R, S as shown in Fig. 10.1 we have:

$$\text{At P} \quad u_P = \tfrac{1}{4}(u_R + u_Q + 0 + \tfrac{2}{9}), \tag{14}$$

$$\text{At Q} \quad u_Q = \tfrac{1}{4}(u_P + u_S + 0 + \tfrac{2}{9}), \tag{15}$$

$$\text{At R} \quad u_R = \tfrac{1}{4}(u_P + u_S + 0 + 0), \tag{16}$$

$$\text{At S} \quad u_S = \tfrac{1}{4}(u_R + u_Q + 0 + 0), \tag{17}$$

where $u_P = u_{1,1}$, $u_Q = u_{2,1}$, $u_R = u_{1,2}$ and $u_S = u_{2,2}$, and the values $\frac{2}{9}$

are obtained from the given boundary conditions along the x-axis (see (8)) by inserting $x = \frac{1}{3}$ and $x = \frac{2}{3}$.

Equations (14)–(17) may be written as

$$4u_P - u_Q - u_R + 0u_S = \tfrac{2}{9}, \quad -u_P + 0u_Q + 4u_R - u_S = 0,$$
$$-u_P + 4u_Q - 0u_R - u_S = \tfrac{2}{9}, \quad 0u_P - u_R - u_Q + 4u_S = 0,$$

$$(18)$$

or in matrix form as

$$\begin{pmatrix} 4 & -1 & -1 & 0 \\ -1 & 4 & 0 & -1 \\ -1 & 0 & 4 & -1 \\ 0 & -1 & -1 & 4 \end{pmatrix} \begin{pmatrix} u_P \\ u_Q \\ u_R \\ u_S \end{pmatrix} = \frac{2}{9} \begin{pmatrix} 1 \\ 1 \\ 0 \\ 0 \end{pmatrix}. \tag{19}$$

By matrix inversion (or any other suitable method) we find the solutions

$$u_P = \tfrac{1}{12}, \quad u_Q = \tfrac{1}{12},$$

$$(20)$$

$$u_R = \tfrac{1}{36}, \quad u_S = \tfrac{1}{36}.$$

These results are necessarily very approximate since the mesh is so coarse. Nevertheless it is of interest to compare these values with those which can be obtained analytically using Fourier's method (see Example 3, Chapter 4). Using equation (77) of Chapter 4 and inserting the boundary condition $f(x) = x(1 - x)$ (from (8)) we have, with $a = b = 1$,

$$u(x, y) = \sum_{r=1}^{\infty} \left\{ \left[2 \int_0^1 x'(1 - x')\sin r\pi x' \, dx' \right] \right.$$
$$\left. \frac{\sin r\pi x \sinh r\pi (1 - y)}{\sinh r\pi} \right\}. \tag{21}$$

Integration by parts gives

$$\int_0^1 x'(1 - x')\sin r\pi x' \, dx' = \frac{2}{r^3 \pi^3}[1 - (-1)^r] \tag{22}$$

whence

144

$$u(x, y) = 2 \sum_{r=1}^{\infty} \frac{2}{r^3 \pi^3} [1 - (-1)^r] \frac{\sin r\pi x \sinh r\pi(1 - y)}{\sinh r\pi} \quad (23)$$

$$= \frac{4}{\pi^3} \sum_{\text{odd } r}^{\infty} \frac{2}{r^3} \cdot \frac{\sin r\pi x \sinh r\pi(1 - y)}{\sinh r\pi} \quad (24)$$

or expanding as an infinite series

$$u(x, y) = \frac{8}{\pi^3} \left[\frac{1}{1^3} \frac{\sin \pi x \sinh \pi(1 - y)}{\sinh \pi} + \right.$$

$$\left. \frac{1}{3^3} \frac{\sin 3\pi x \sinh 3\pi(1 - y)}{\sinh 3\pi} + \frac{1}{5^3} \frac{\sin 5\pi x \sinh 5\pi(1 - y)}{\sinh 5\pi} + \dots \right]. \quad (25)$$

Now at the point P, $x = \frac{1}{3}$, $y = \frac{1}{3}$ so

$$u(\tfrac{1}{3}, \tfrac{1}{3}) = \frac{8}{\pi^3} \left[\frac{\sin \dfrac{\pi}{3} \sinh \dfrac{2\pi}{3}}{\sinh \pi} + \frac{1}{27} \frac{\sin \pi \sinh 2\pi}{\sinh 3\pi} + \right.$$

$$\left. \frac{1}{125} \frac{\sin \dfrac{5\pi}{3} \sinh \dfrac{10\pi}{3}}{\sinh 5\pi} + \dots \right]. \quad (26)$$

The second term is zero since $\sin \pi = 0$, and the third term is small compared to the first. Accordingly

$$u(\tfrac{1}{3}, \tfrac{1}{3}) \approx \frac{8}{31.006} \left[\left(\frac{\sqrt{3}}{2} \right) \frac{3.998}{11.549} \right] = 0.077, \quad (27)$$

which is to be compared with the value of $\frac{1}{12}$ ($= 0.083$) obtained from the finite difference method.

It should be clear from this example that the solution of partial differential equations using the finite difference methods leads to sets of algebraic equations expressible in matrix form as $\mathbf{AU} = \mathbf{B}$, where $\mathbf{U}$ is a column vector of unknowns at the interior nodes and $\mathbf{B}$ is a column vector of values of $u(x, y)$ at the boundary nodes. With a coarse mesh it is impossible to map an irregular-shaped boundary at all accurately, but with a finer mesh the accuracy can clearly be much improved (see Fig. 10.2). The nearest nodes to the boundary are taken to be those at which the boundary values are given. Taking a finer mesh necessarily increases the size of the matrices, and for large systems a computer program is best used for the inversion of the matrix of coefficients **A**. It is plausible that using a finer mesh may, besides allowing a more accurate mapping of the boundary, improve the accuracy of the final solution. However, this is not always so. In certain cases the numerical errors can accumulate and the solution become unstable as the mesh size is decreased. This is a complicated and very important problem and is dealt with more fully in the standard texts listed under Further Reading.

Another difficult point which frequently arises is the use of the finite difference method near a singular point. The validity of the method depends on the approximations for the derivatives given by (3), (4) and (5) being reasonably good. Near a singular point, where the function and its derivatives can become increasingly large, this is clearly not the case. In such instances, it is usually best to examine the solution near the singularity by analytic methods and then to

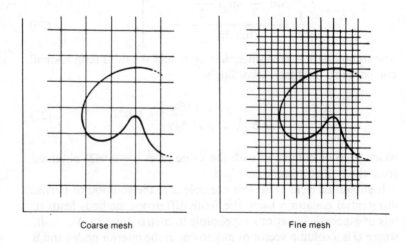

Coarse mesh   Fine mesh

Fig. 10.2.

join on to a numerical solution some distance away where the numerical method is a reasonably good approximation (see [15], [18] in Further Reading).

We now conclude this section with an example relating to the one-dimensional heat conduction (or diffusion) equation.

Example 2. Solve the equation

$$\frac{\partial^2 u}{\partial x^2} = \frac{1}{k}\frac{\partial u}{\partial t} \tag{28}$$

using the finite difference method given the boundary conditions

$$u(0, t) = u(1, t) = 0, t \geqslant 0 \text{ and } u(x, 0) = x(1 - x), 0 \leqslant x \leqslant 1. \tag{29}$$

This problem can be solved analytically (see Example 1, Chapter 4). Taking $f(x) = x(1 - x)$ and $l = 1$, we have (from Chapter 4, equation (48))

$$u(x, t) = \sum_{r=1}^{\infty} \left[2\int_0^1 x'(1 - x')\sin r\pi x' \, dx' \right] e^{-r^2\pi^2 kt}\sin r\pi x. \tag{30}$$

The integral $\int_0^1 x'(1 - x')\sin r\pi x' \, dx'$ has already been evaluated in

(22) and has the value $\frac{2}{r^3\pi^3}[1 - (-1)^r]$. Hence

$$u(x, t) = \frac{4}{\pi^3}\sum_{r=1}^{\infty}\frac{1}{r^3}[1 - (-1)^r]e^{-r^2\pi^2 kt}\sin r\pi x \tag{31}$$

$$= \frac{8}{\pi^3}\sum_{\text{odd } r}\frac{1}{r^3} \cdot e^{-r^2\pi^2 kt}\sin r\pi x. \tag{32}$$

Now for the purposes of a comparison with the following calculations using the finite difference method we take the constant k in the equation equal to unity. Using the results given in (3) and (6) the equation may be approximated by

147

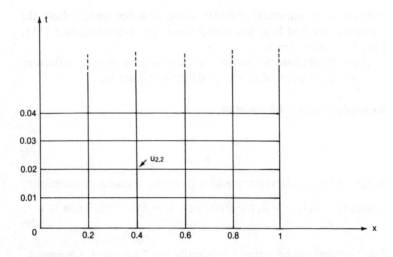

Fig. 10.3.

$$\frac{u_{m,n+1} - u_{m,n}}{g} = \frac{u_{m+1,n} - 2u_{m,n} + u_{m-1,n}}{h^2}, \tag{33}$$

where g is the mesh length in the t-direction and h is the mesh size in the x-direction. Suppose we take $h = 0.2$ and $g = 0.01$ (see Fig. 10.3). Then

$$\begin{aligned}
u_{m,n+1} &= \tfrac{1}{4}(u_{m+1,n} - 2u_{m,n} + u_{m-1,n}) + u_{m,n} \\
&= \tfrac{1}{4}(u_{m+1,n} + 2u_{m,n} + u_{m-1,n}).
\end{aligned} \tag{34}$$

Using the boundary condition $u = x(1 - x)$ for $0 \leqslant x \leqslant 1$, $t = 0$, we have

$$u_{0,0} = u(0, 0) = 0, \quad u_{1,0} = u(0.2, 0) = \frac{4}{25},$$

$$u_{2,0} = u(0.4, 0) = \frac{6}{25}, u_{3,0} = u(0.6, 0) = \frac{6}{25}, \tag{35}$$

$$u_{4,0} = u(0.8, 0) = \frac{4}{25}, u_{5,0} = u(1, 0) = 0.$$

148

These values together with (34) can now be used to generate the values $u_{1,1}$, $u_{2,1}$, $u_{3,1}$, $u_{4,1}$ as follows:

$$u_{1,1} = \frac{1}{4}(u_{2,0} + 2u_{1,0} + u_{0,0}) \quad = \frac{1}{4}\left(\frac{6}{25} + \frac{8}{25} + 0\right) \quad = \frac{7}{50}, \quad (36)$$

$$u_{2,1} = \frac{1}{4}(u_{3,0} + 2u_{2,0} + u_{1,0}) = \frac{1}{4}\left(\frac{6}{25} + \frac{12}{25} + \frac{4}{25}\right) = \frac{11}{50}, \quad (37)$$

$$u_{3,1} = \frac{1}{4}(u_{4,0} + 2u_{3,0} + u_{2,0}) = \frac{1}{4}\left(\frac{4}{25} + \frac{12}{25} + \frac{6}{25}\right) = \frac{11}{50}, \quad (38)$$

$$u_{4,1} = \frac{1}{4}(u_{5,0} + 2u_{4,0} + u_{3,0}) = \frac{1}{4}\left(0 + \frac{8}{25} + \frac{6}{25}\right) \quad = \frac{7}{50}, \quad (39)$$

making use of the boundary conditions $u_{0,0} = 0$ and $u_{5,0} = 0$. Likewise we can now make use of these values to generate the next set of values to get

$$u_{1,2} = \frac{1}{4}(u_{2,1} + 2u_{1,1} + u_{0,1}) \quad = \frac{1}{4}\left(\frac{11}{50} + \frac{14}{50} + 0\right) \quad = \frac{1}{8}, \quad (40)$$

$$u_{2,2} = \frac{1}{4}(u_{3,1} + 2u_{2,1} + u_{1,1}) = \frac{1}{4}\left(\frac{11}{50} + \frac{22}{50} + \frac{7}{50}\right) = \frac{1}{5}, \quad (41)$$

$$u_{3,2} = \frac{1}{4}(u_{4,1} + 2u_{3,1} + u_{2,1}) = \frac{1}{4}\left(\frac{7}{50} + \frac{22}{50} + \frac{11}{50}\right) = \frac{1}{5}, \quad (42)$$

$$u_{4,2} = \frac{1}{4}(u_{5,1} + 2u_{4,1} + u_{3,1}) \quad = \frac{1}{4}\left(0 + \frac{14}{50} + \frac{11}{50}\right) \quad = \frac{1}{8}, \quad (43)$$

using the boundary conditions $u_{0,1} = 0$ and $u_{5,1} = 0$.
We now compare the numerical solution $u_{2,2}$ with the exact solution (32) evaluated at $x = 0.4$ ($m = 2$) and $t = 0.02$ ($n = 2$) given by

$$u(0.4, 0.02) = \frac{8}{\pi^3}\left[\frac{1}{1^3}e^{-0.02\pi^2}\sin(0.4\pi) + \right.$$

$$\left. \frac{1}{3^3}e^{-9(0.02)\pi^2}\sin(1.2\pi) + \ldots\right] \quad (44)$$

which gives

$$u(0.4, 0.02) \simeq \frac{8}{\pi^3}(0.781 - 0.004) = 0.200. \qquad (45)$$

The numerical value of $u_{2,2}$ is seen to be identical to the analytically calculated value to a high degree of accuracy.

10.3 Variational calculus – the finite element method

The simplest problem of the branch of mathematics known as the calculus of variations (or variational calculus) is to determine the function $y(x)$ which makes the integral

$$I[y] = \int_a^b f(x, y(x), y'(x)) \, dx \qquad (46)$$

stationary, where a and b are given limits of integration, f is some specified function, and $y'(x) = \dfrac{dy}{dx}$. The notation $I[y]$ denotes that the integral is a function of the function $y(x)$, and is termed a *functional*. In many cases the stationary value will be a minimum value and in others a maximum. Fairly complicated tests are known which determine the precise nature of the stationary value, but in most applications it is possible to verify directly whether or not a minimum value has been obtained. It can be proved that a stationary value of (46) is attained if $y(x)$ is the solution of the Euler–Lagrange equation

$$\frac{\partial f}{\partial y} - \frac{d}{dx}\left(\frac{\partial f}{\partial y'}\right) = 0. \qquad (47)$$

For a specified function f, (47) becomes an ordinary differential equation for $y(x)$.

Suppose now we wish to solve a differential equation which is the Euler–Lagrange equation associated with some $f(x, y, y')$. The solution of the differential equation is then the function $y(x)$ which makes (46) stationary.

Consider, for example, the equation

$$y'' + y + x = 0 \qquad (48)$$

subject to the boundary conditions $y(0) = 0$, $y(1) = 0$. This equation is easily seen to be the Euler–Lagrange equation associated with the function

$$f(x, y, y') = (y')^2 - y^2 - 2xy \tag{49}$$

for, since

$$\frac{\partial f}{\partial y} = -2y - 2x, \quad \frac{\partial f}{\partial y'} = 2y', \tag{50}$$

we have by (47) and (50)

$$\frac{\partial f}{\partial y} - \frac{d}{dx}\left(\frac{\partial f}{\partial y'}\right) = -2y - 2x - \frac{d}{dx}(2y') = -2(y'' + y + x) = 0, \tag{51}$$

which is (48).

The solution of (48) therefore is that function $y(x)$ which makes the integral (or functional)

$$I[y] = \int_0^1 \{(y')^2 - y^2 - 2xy\}\,dx \tag{52}$$

stationary. If we now insert into (52) a trial solution which satisfies the boundary conditions but which contains a finite number of arbitrary constants $c_1, c_2, \ldots, c_n$ (say), then I will be a function of these n constants and, in the usual way, will be stationary for those values of the constants which are solutions of the equations

$$\frac{\partial I}{\partial c_1} = 0, \quad \frac{\partial I}{\partial c_2} = 0, \ldots, \quad \frac{\partial I}{\partial c_n} = 0. \tag{53}$$

For example, if we choose

$$y_1(x) = c_1 x(1 - x) \tag{54}$$

as a possible trial function satisfying the given boundary conditions $y(0) = 0$, $y(1) = 0$ imposed on (48), then putting (54) into (52) we find after integrating that

$$I = \frac{3}{10}c_1^2 - \frac{1}{6}c_1. \tag{55}$$

Hence

$$\frac{\partial I}{\partial c_1} = \frac{3}{5}c_1 - \frac{1}{6} \tag{56}$$

151

which gives

$$c_1 = \frac{5}{18}. \tag{57}$$

Accordingly the first approximation to the solution of (48) satisfying the given boundary conditions is

$$y_1(x) = \frac{5}{18}x(1 - x). \tag{58}$$

Better approximations can be obtained by adopting

$$y_n(x) = x(1 - x)(c_1 + c_2 x + \ldots + c_n x^{n-1}) \tag{59}$$

as a trial solution, and solving for the $c_1, c_2, \ldots, c_n$ in the same way. This method, known as the Rayleigh–Ritz method, is relatively simple when the number of constants, n, is small, but leads to a large amount of algebra when n is large. In these circumstances, it is sometimes better to divide the range of integration into a number of sub-ranges (or finite elements) and to use a simple trial function over each sub-range (care being taken to ensure continuity of these trial functions between each finite element and the next). The total integral over the basic range is then just the sum of the integrals over each element. We now consider a simple example of this finite element method.

Example 3. The equation (48) with the specified boundary conditions $y(0) = 0$, $y(1) = 0$ is the Euler–Lagrange equation associated with the functional

$$I[y] = \int_0^1 \{(y')^2 - y^2 - 2xy\} dx, \tag{60}$$

as we have already seen. We now wish to minimise this integral using the finite element approach. To do this we divide the range $0 \leqslant x \leqslant 1$ into two elements of equal lengths ($\frac{1}{2}$) and take the first approximation as

$$y_1(x) = \begin{cases} c_1 x, & 0 \leqslant x \leqslant \frac{1}{2}, \\ c_1(1 - x), & \frac{1}{2} \leqslant x \leqslant 1, \end{cases} \tag{61}$$

152

so that $y_1(x)$ is continuous at $x = \frac{1}{2}$, and satisfies the boundary conditions at $x = 0$ and $x = 1$.
Now

$$\int_0^{1/2} \{(y')^2 - y^2 - 2xy\}\,dx = \int_0^{1/2} (c_1^2 - c_1^2 x^2 - 2c_1 x^2)\,dx$$

$$= \frac{c_1^2}{2} - \frac{c_1^2}{24} - \frac{c_1}{12}, \qquad (62)$$

and

$$\int_{1/2}^1 \{(y')^2 - y^2 - 2xy\}\,dx = \int_{1/2}^1 (c_1^2 - c_1^2(1-x)^2 - 2c_1 x(1-x))\,dx$$

$$= \frac{c_1^2}{2} - \frac{c_1^2}{24} - \frac{c_1}{6}. \qquad (63)$$

Hence

$$I[y] = \frac{11}{12}c_1^2 - \frac{c_1}{4} \qquad (64)$$

which, by differentiation with respect to c_1, has a stationary value when $c_1 = \dfrac{3}{22}$. Hence the first approximation to the solution of (48) is

$$y_1(x) = \begin{cases} \dfrac{3}{22}x, & 0 \leqslant x \leqslant \frac{1}{2}, \\[2mm] \dfrac{3}{22}(1-x), & \frac{1}{2} \leqslant x \leqslant 1, \end{cases} \qquad (65)$$

which can be compared with the first approximation (54) obtained by the Rayleigh–Ritz method.

By dividing the range of integration into three finite elements (of equal length $1/3$) we could construct a second approximation in the form

153

$$y_2(x) = \begin{cases} c_1 x, & 0 \leqslant x \leqslant \frac{1}{3}, \\ c_1\left(\frac{2}{3} - x\right) + c_2\left(x - \frac{1}{3}\right), & \frac{1}{3} \leqslant x \leqslant \frac{2}{3}, \\ c_2(1 - x), & \frac{2}{3} \leqslant x \leqslant 1, \end{cases} \quad (66)$$

noting again that continuity has been preserved at the joins of the finite elements, and that the boundary conditions are also satisfied. Proceeding as before, by summing the integrals over the three elements and differentiating with respect to c_1 and c_2, the solution may be found.

The calculus of variations and the finite element method may both be readily extended to functions of two variables with the consequence that the finite element method especially may be applied in the numerical solution of certain types of *partial* differential equations. This method has, in fact, attracted much attention amongst engineers in recent years and finite element computer software packages are readily available for a variety of problems. It is not possible in this short chapter to deal with these extensions in detail, but the following short account may indicate the general approach. A fundamental result from the calculus of variations is that the solution of Laplace's equation

$$\frac{\partial^2 u}{\partial x^2} + \frac{\partial^2 u}{\partial y^2} = 0 \quad (67)$$

within a region R with prescribed (Dirichelet) boundary conditions on the bounding curve of R is the function $u(x, y)$ which minimises the integral

$$I[u] = \frac{1}{2} \iint_R \left\{ \left(\frac{\partial u}{\partial x}\right)^2 + \left(\frac{\partial u}{\partial y}\right)^2 \right\} dx dy \quad (68)$$

subject to the same boundary conditions. In fact, Laplace's equation is just the Euler–Lagrange equation of the functional (68) (see [14]). The finite element method now consists in dividing the region R into a number of smaller sub-regions (or finite elements). Frequently a division into triangular regions is taken as these can very easily be made to fit inside an irregular-shaped region, especially since the size of the triangles can be varied as required. Over each of these sub-regions some trial function is used, continuity between finite elements again being demanded, and the trial functions over the triangles bordering the boundary being so chosen that the given boundary conditions are satisfied. The integral over the region R is then the sum of the integrals over all the sub-regions. The arbitrary constants in the trial functions are determined in the usual way by differentiating the resulting integral with respect to these constants in turn and solving the resulting set of equations.

This powerful method is well-explained in [19].

FURTHER READING

1. SNEDDON, I. N., *Elements of Partial Differential Equations*, McGraw-Hill 1957.
2. DUFF, G. F. D. and NAYLOR, D., *Differential Equations of Applied Mathematics*, Wiley 1966.
3. SAGAN, H., *Boundary and Eigenvalue Problems in Mathematical Physics*, Wiley 1961.
4. SOMMERFELD, A., *Partial Differential Equations in Physics*, Academic Press 1949.
5. TYCHONOV, A. N. and SAMARSKII, A. A., *Equations of Mathematical Physics*, Pergamon Press 1963.
6. MACKIE, A. G., *Boundary Value Problems*, Oliver and Boyd 1965.
7. SNEDDON, I. N., *Special Functions of Mathematical Physics and Chemistry*, Oliver and Boyd 1956.
8. TRANTER, C. J., *Integral Transforms in Mathematical Physics*, Methuen 1956.
9. SNEDDON, I. N., *Fourier Transforms*, McGraw-Hill 1951.
10. SMITH, M. G., *Laplace Transform Theory*, Van Nostrand 1966.
11. CHURCHILL, R. V., *Complex Variables and Applications*, McGraw-Hill 1960.
12. COURANT, R. and HILBERT, D., *Methods of Mathematical Physics*, Vols. 1 and 2, Wiley 1961.
13. MORSE, P. M. and FESHBACH, H., *Methods of Theoretical Physics*, Vols. 1 and 2, McGraw-Hill 1953.
14. WEINSTOCK, R., *Calculus of Variations with Applications to Physics and Engineering*, McGraw-Hill 1952.
15. SMITH, G. D., *Numerical Solution of Partial Differential Equations*, Clarendon Press (2nd edn) 1978.
16. SMITH, M. G., *Theory of Partial Differential Equations*, Van Nostrand 1967.
17. LIGHTHILL, M. J., *Fourier Analysis and Generalised Functions*, Cambridge University Press 1964.
18. AMES, W. F., *Numerical Methods for the Solution of Partial Differential Equations*, Nelson (2nd edn) 1977.
19. MITCHELL, A. R. and WAIT, R., *The Finite Element Method in Partial Differential Equations*, Wiley 1977.

ANSWERS TO PROBLEMS

PROBLEMS 1

1. (a) $\dfrac{\partial u}{\partial x} - \dfrac{\partial u}{\partial y} = 0$, (b) $x\dfrac{\partial u}{\partial x} - y\dfrac{\partial u}{\partial y} = 0$,

   (c) $\dfrac{\partial^2 u}{\partial x^2} - \dfrac{\partial^2 u}{\partial y^2} = 0$, (d) $x\dfrac{\partial u}{\partial x} + y\dfrac{\partial u}{\partial y} = nu$.

PROBLEMS 2

1. (a) Hyperbolic; $u = f[x + (2 + \sqrt{7})y] + g[x + (2 - \sqrt{7})y]$.
   (b) Parabolic; $u = f(x + y) + xg(x + y)$.
   (c) Elliptic; $u = f(x + 2iy) + g(x - 2iy)$.
   (d) Parabolic; $u = f(x - \tfrac{1}{2}y) + xg(x - \tfrac{1}{2}y)$.
   (e) Elliptic; $u = f[x + i\sqrt{(2)}y] + g[x - i\sqrt{(2)}y]$.

3. (a) $u = \dfrac{1}{2c}[\tan^{-1}(x + ct) - \tan^{-1}(x - ct)]$.

   (b) $u = \dfrac{x^3}{3}y + xy + y^2$.

   (c) $u = xy + \tfrac{1}{2}(x^2 + y^2)$.

PROBLEMS 3

2. (a) $\dfrac{1}{\sqrt{\pi}}$, $\sqrt{\left(\dfrac{2}{\pi}\right)}\cos x$, $\sqrt{\left(\dfrac{2}{\pi}\right)}\cos 2x, \ldots$

   (b) $\sin \pi x$, $\sin 2\pi x$, $\sin 3\pi x, \ldots$
3. $\alpha = 0$, $\beta = -3$.

PROBLEMS 4

1. $u = -2\displaystyle\sum_{n=1}^{\infty}\dfrac{(-1)^n}{n}\,e^{(1+n^2)(1-y)}\sin nx$.

PROBLEMS 7

2. (a) $-\sin x + \sqrt{2}\sin(x\sqrt{2})$.
   (b) $-1 + e^x(1 - x) + \tfrac{1}{2}x^2 e^x$.
   (c) $H(x - 1)e^{-(x-1)}$.
   (d) $1 - H(x - 1)$.

5. $u = \tfrac{1}{2}x + \tfrac{3}{4}\sin 2x$.

158

6. (a) $y = e^{-x} \cos x$.

   (b) $y = e^{-x} + \int_0^x f(\xi) e^{-(x-\xi)} d\xi$.

PROBLEMS 8

2. $u = U_1\left(1 - \dfrac{x}{l}\right) + \dfrac{x}{l} U_2 + \dfrac{2}{\pi} \sum_{n=1}^{\infty} \dfrac{1}{n} (U_2 \cos n\pi - U_1) e^{-n^2 \pi^2 kt/l^2} \sin \dfrac{n\pi x}{l}$.

3. $u = \dfrac{4C}{\pi^{3/2}} \sum_{n=1}^{\infty} \left(\dfrac{1 - \cos n\pi}{n}\right) \sin nx \int_{y/2\sqrt{kt}}^{\infty} e^{-(p^2 + n^2 y^2/4p^2)} dp$.

INDEX

Index